T5-ARZ-486

CULTURE AND CURRENCY

POLITICAL CULTURES

Aaron Wildavsky, Series Editor

Political cultures broadly describe people who share values, beliefs, and preferences legitimating different ways of life. This series will be distinguished by its openness to a variety of approaches to the study of political cultures; any defensible comparison, definition, and research method will be considered. The goal of this series is to advance the study of political cultures conceived generally as rival modes of organizing political and social life.

A single set of common concerns will be addressed by all authors in the series: what values are shared, what sorts of social relations are preferred, what kinds of beliefs are involved, and what the political implications of these values, beliefs, and relations are. Beyond that, the focal points of the studies are open and may compare cultures within a country or among different countries, including or excluding the United States.

Books in the Series

Culture and Currency: Cultural Bias in Monetary Theory and Policy
John W. Houghton

A Genealogy of Political Culture
Michael E. Brint

Cultural Theory
Michael Thompson, Richard Ellis, and Aaron Wildavsky

District Leaders: A Political Ethnography
Rachel Sady

The American Mosaic: The Impact of Space, Time, and Culture on American Politics
Daniel J. Elazar

CULTURE AND CURRENCY

Cultural Bias in Monetary Theory and Policy

John W. Houghton

Westview Press
BOULDER, SAN FRANCISCO, & OXFORD

Political Cultures

This Westview softcover edition is printed on acid-free paper and bound in library-quality, coated covers that carry the highest rating of the National Association of State Textbook Administrators, in consultation with the Association of American Publishers and the Book Manufacturers' Institute.

Published in 1991 in the United States of America by Westview Press, Inc., 5500 Central Avenue, Boulder, Colorado 80301-2847, and in the United Kingdom by Westview Press, 36 Lonsdale Road, Summertown, Oxford OX2 7EW

Library of Congress Cataloging-in-Publication Data
Houghton, John W.
Culture and currency : cultural bias in monetary theory and policy / John W. Houghton. — Westview softcover ed.
p. cm. — (Political cultures)
Includes bibliographical references and index.
ISBN 0-8133-1191-8
1. Currency question—Great Britain—History—19th century.
2. Precious metals—Great Britain—History—19th century.
3. Monetary policy—Great Britain—History—19th century.
4. Monetary policy—Social aspects. I. Title. II. Series.
HG939.H76 1991
332.4′941—dc20 91-25650
CIP

Printed and bound in the United States of America

The paper used in this publication meets the requirements of the American National Standard for Permanence of Paper for Printed Library Materials Z39.48-1984.

10 9 8 7 6 5 4 3 2 1

Contents

CHAPTER 1

Introduction

*If I were asked to say briefly and superficially what I regard as the linchpin of the reformation [of science] to which I have alluded, I would say it was the rethinking of our theory of knowledge in terms of evolving figurations of people, of developing groups of independent individuals as the subject of knowledge rather than of an isolated individual of the homo clausus type.**

The aim of this book is to shed light on how people come to hold opposing views, how these views solidify into the sides of a debate and how one side becomes the dominant view.

> Why, as all have access to the same nature, physical and human, don't they come to the same conclusions? Or, if each individual is different, why don't they come to wholly different conclusions? A sociology of perception must explain both why the world resembles neither an epistemological Tower of Babel in which communication between individuals is impossible nor a homogenized blend in which communication is no longer necessary.[1]

At one level the subject is choice or decision making, but underlying this it is meaning. People come to hold the views they do out of the process of making sense of their lives and the world in which they live. Our ideas about how the world works derive from our social relations. Individuals internalize the social order for a purpose. That purpose is to sustain a way of life. Only by recognizing the social origins of thought can we account for both the shared nature of categories and their variation across different places and times.[2]

These socially constructed cultural biases inform the decision process, and thus profoundly affect choice. The real moment of choosing is

---

N. Elias, H. Martins, and R. Whitley, Eds., 1982, *Scientific Establishments and Hierarchies,* Sociology of the Sciences Year Book (Dordrecht: Reidel), p. 62.

choice of comrades and their way of life, and from this choice about how to relate to other people are derived the myriad preferences that make up everyday life.[3] This is the basis of the cultural theory approach that we adopt. The aims are to investigate the generation of categories and the development of shared symbolic schemas or theories that seek to explain the workings of the world, and to explain theory choice.

Philosophical approaches to theory choice in science are a grounding from which to begin our search for an explanation of the patterned nature of sense making. However, traditional philosophical approaches seem rather partial and inclined to be misleading. Put in the very simplest terms there are two basic approaches to the problem in the history and philosophy of science. The first is a mainstream basket of approaches of the positivist/empiricist ilk. The underlying theme of this type of approach is that there is one (empirical) reality out there, and the history of the sciences is a progression toward coming to know it. There is one correct knowing, one correct meaning to be gained, and any alternatives are merely erroneous. Second, a mainstream alternative to this view consisting of a raft of approaches that are variously idealist/rationalist. The underlying theme of these approaches is that the world appears to people as it does because of some quality within them. This quality might be anything from the rational structure of the human mind—shared by all humans as a consequence of their physiology, or by larger or smaller proportions of humankind as a consequence of their socialization—to the uniqueness of each individual.

The result of these two broad approaches is that knowledge, meaning or choice is seen as either absolute (right or wrong) or relative (dependent upon the individual). The latter is said to lead to a state wherein any view, any interpretation, any sense, is as good as any other. Hence the problem polarizes into the extreme views that either "there is no alternative" or any knowledge is as good as any other—"it is all relative." It is this stark choice that leaves us with the questions we have posed. How do people come to hold opposing views about the same thing, how do the sides of a debate solidify into sides, and how does one side emerge as the dominant paradigm?

The key with which we seek to unlock the apparent dilemma is to see action and knowing or meaning as dialectically inter-related. That is, to see "making sense of what one does" and "doing what makes sense" as the essential conditions of life—in the long run. Hence we are looking for matches between patterns of acting and patterns of thinking. The underlying notion is that meaning attribution is a symbolization process, wherein nature and society are understood in terms of a schema or theory that is a symbolic (re)description of life in society and/or nature. Following Durkheim we hypothesize that the structure of symbolism parallels the structure of social life. Hence our focus is on the patterned inter-relation of individuals' social environments and their symbolic re-descriptions, schemas or theories.

The first hurdle in outlining a cultural theory is that of terminology, and the many and various definitions of the slippery notion *culture.* Rather than add to the confusion and extend semantic debate, the practitioners of cultural theory adopt three key terms that replace three of the main ways that the term *culture* is used. First, *cultural bias* refers to the shared values, beliefs, norms, rationalizations, symbols and ideologies that are the basis of mental products. Cultural biases include the taken-for-granted, the self-evident, the heuristics of decision making and the automatic pilots of life. Second, *social relations* refers to the patterns of inter-personal relations, the structure of social relations and social organization. Third, *way of life* refers to the viable, sustainable combinations of social relations and cultural bias.[4] As E. E. Schattschneider said: "Organization is the mobilization of bias."[5] Organized social life is founded on the mobilization of cultural bias.

The case study that is explored at length in this book involves both the development of a theoretical approach, and formulation of policy based on that theory, together with debate on alternative theoretical approaches and policy options. Hence we explore the questions of theory choice in the history of ideas/history of science, and issues—including theoretical or expert knowledge—relating to public policy selection and implementation. This book is, therefore, of much wider interest than that of its substantive case study alone. It is of importance to anyone interested in knowledge, in the history and philosophy of science, history of ideas and of thought, disciplinary history, the study of belief systems, etc.; anyone interested in attitudes, in political sociology, ideology, political cultures, etc.; and anyone interested in the social practices of elites, elite studies, elite cultures, etc. It is a study that undertakes the development and operationalization of a cultural theory approach, which promises much for the study of knowledge and knowledge claims.

What our cultural theory-based approach involves is in essence an explanation of the nature of choice based on the structure of interpersonal relations and on cultural bias. It is an alternative to the mainstream dualism. It is neither an empiricist or rationalist absolute view in which "there is no alternative," nor an idealist or structuralist relative view in which "it is all relative." It is, rather, a cultural view in which there is a five-way pluralism. Underlying this posited pluralism is the notion that traditional social science approaches based on either societies or individuals incorrectly exclude the middle ground, communities. A focus on communities allows cultural theorists to overcome the limitation of the stark choice between absolutism and relativism, and to see a pluralism composed of five sustainable ways of life.

Believing that "the proof of the pudding is in the eating," the major element of this book is a case study undertaken from the perspective outlined in the next chapter. Indeed this work is a demonstration, rather than merely an exhortation. Our case study encompasses a period during which the very foundations of monetary theory, of economics and of economic and political discourse, were being laid—a period that is at the foundation of modern thinking itself, characterized by the revolutionary changes, economic, social and political, that were the industrial revolution. The inter-relation of currency and culture in that key period is examined.

By way of an outline, the first task undertaken is to develop a practicable means of analyzing particular theories and ideas from the history of science or history of knowledge. To this end a three-dimensional cultural theory schema is developed. This three-dimensional model allows the development of a methodological tool which relates styles or modes of thought, and thereby the developing analytical content of theory, to the social relations experienced in the social environment in terms of grid and group dimensions. It also locates the dynamic of knowledge in the social actions—power relations—that animate the reproduction and transformation of the real underlying structure of social relations in societies. A key feature of this schema is its incorporation of insights derived from catastrophe theory which permit the formal handling of both gradual and sudden change, of evolution and revolution.

The suggested methodological approach is applied, by way of example, to the case of the development of bullionist monetary theory and policy in England at the turn of the nineteenth-century. It is a story of the stylistic and substantive polarization of monetary thought into bullionist and anti-bullionist camps, and of the faltering passage of bullionist theory into policy practice during the period 1797 to 1819–21. The story is structured in terms of the four major phases of the debate: first, the emergence of bullionism in the context of the debate over Irish currency circa 1803; second, the development of bullionism in the context of the English currency debate, which culminated in the *Bullion Report* of 1810; third, the reception of bullionist ideas in the context of the failure of bullionist principles to gain sway as policy practice circa 1811; and fourth, the adoption and application of those ideas as policy practice in the form of *Peel's Bill* circa 1819–21.

In each of these chapters a brief historical sketch is analyzed in the light of the proposed schema as a means of explaining the broad phases of and the specific developments within the ongoing debate. The ways in which the suggested approach outperforms conventional approaches is revealed. Whereas both positivist and idealist approaches have left knowledge production and change out of picture—in the first case through treating it as an intellectual process independent of practice, and in the second through treating it as a psychological mystery—the suggested approach involves a truly social (cultural) theory of knowledge, one that recognizes the inter-relation between interpretation of, and intervention in, the world.

While the substantive topic, early nineteenth-century monetary theory, may seem an obscure and perhaps rather difficult one, we assure the reader that it makes a fascinating story and gives rewards far beyond those relating to monetary theory and policy, or even economic thought as such. The aim is not to engage in an exegetical analysis of obscure and aged texts, but rather to develop and demonstrate a cultural theory of knowledge. It is, therefore, a study for all who are interested in knowledge, ideas, ideology, elites, political cultures and the history of the era in which the roots of our own are set.

NOTES

1. M. Thompson, R. Ellis, and A. Wildavsky, 1990, *Cultural Theory* (Boulder, Colorado: Westview Press), p. 129.
2. Thompson, Ellis and Wildavsky, op. cit., citing E. Durkheim and M. Mauss, 1963, *Primitive Classification,* translated by R. Needham (Chicago: University of Chicago Press), and E. Durkheim, 1965, *The Elementary Forms of Religious Life: Study in Religious Sociology,* translated by Joseph Ward Swain (New York: Free Press).
3. Thompson, Ellis and Wildavsky, op. cit., p. 57.
4. Ibid., p. 1.
5. Cited by S. Clegg, 1989, *Frameworks of Power* (London: Sage), p. 76.

CHAPTER 2

Historiography and Cultural Theory

> *At the heart of this method is the concept of a* style of thought. *The history of thought from this point of view is no mere history of ideas, but an analysis of different* styles of thought *as they grow and develop, fuse and disappear; and the key to the understanding of these changes in ideas is to be found in the changing social background, mainly in the fate of the social groups or classes which are the "carriers" of these styles of thought.**

Anthropologists have found that there are in fact a limited number of sustainable ways of life. These depend on matching the way of thinking about or orientation to the world, to the way of living or acting in it, and on the existence of the other ways of life as practical and intellectual sounding boards. This is the basis for our cultural theory or theory of socio-cultural viability. The underlying propositions are: that there are few (five) sustainable ways of life, that each is internally coherent and involves a specific combination of social relations and cultural bias, and that each depends on the coexistence of one or more of the other ways of life.

The founding work in developing this approach was that of Mary Douglas.[1] Douglas sought a taxonomic systematization of forms of social commitment, believing that, despite apparent ethnographic diversity, there were actually few such basic forms. Her aim was to be able to link the underlying convictions that people hold about the world to the forms of social order and the structure of social relations they experience. Her schema aimed to systematically account for the distribution of beliefs in terms of variations in social relations.

Douglas and followers developed a two-dimensional model comprised of the social dimensions of grid and group—the grid/group model. *Group* refers to the extent of incorporation into and commitment to

---

K. Mannheim, 1953, "Conservative Thought," in P. Kecskemeti, Ed., *Essays on Sociology and Social Psychology: By Karl Mannheim* (London: Routledge and Kegan Paul), p. 74.

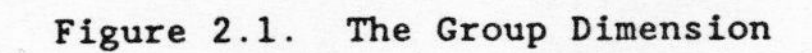
Figure 2.1. The Group Dimension

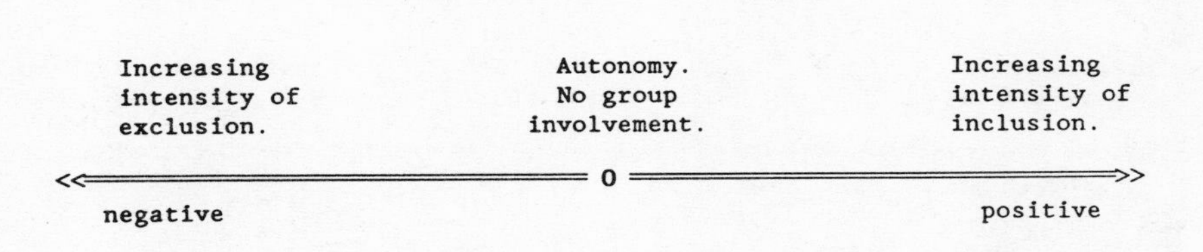

Source: M. Thompson, 1982, "A Three-Dimensional Model," In Mary Douglas, Ed., *Essays in the Sociology of Perception* (London: Routledge and Kegan Paul), p. 40.

an identifiable group, the extent to which an actor's social life depends on membership in a social group or groups. A group's admission rules may be weak or strong, it may be more or less exclusive; a member's life support may be drawn completely or partially from the group,[2] members' individual choices will be more or less subject to group determination. According to these factors an actor can be placed on a continuum between no, and extreme, group dependence. *Grid* refers to the extent of regulation an actor experiences, that is, the degree to which an actor's actions are controlled or regulated, the extent to which an actor's social life is restricted by preordained and imposed rules. Actors are positioned somewhere between totally regulated, and totally free from such regulation; between no, and total individual negotiation.

Group is a measure of the extent to which an individual's life is absorbed in and sustained by group membership. An individual joined with others in common residence, shared work, shared resources and recreation would be assigned a high group position. The higher the group boundaries, and the tighter its rules of admission and membership, the higher the group position.[3] The group dimension is an expression of the extent of group involvement and commitment. It embraces both positive and negative involvement; inclusion and exclusion respectively. (See Figure 2.1.)

Grid is an expression of the extent of social regulation. When individuals are free to negotiate their relationships with one another they are in a low grid environment, whereas in a high grid environment there are institutionalized classifications that regulate interaction and keep individuals apart. In such a situation there are predefined social positions for actors, and they are not free to re-negotiate their lot.[4] Grid too has both negative and positive extension, regulating and regulated respectively.[5] (See Figure 2.2.)

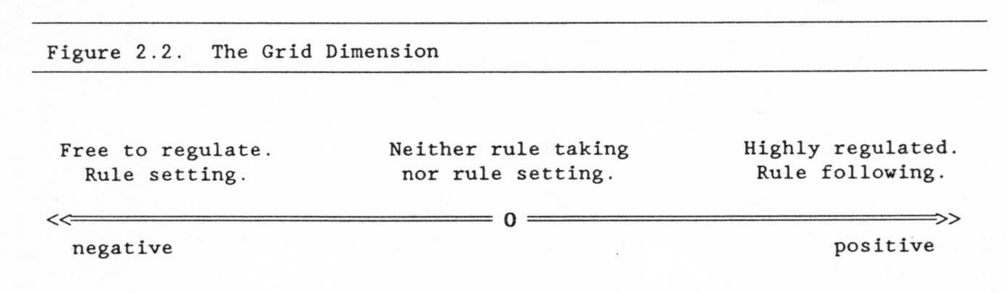

Figure 2.2. The Grid Dimension

Free to regulate. Rule setting.	Neither rule taking nor rule setting.	Highly regulated. Rule following.
<<═══════════	0	═══════════>>
negative		positive

The grid and group dimensions are commonly drawn as x,y axes, and the defined space divided into quadrants which represent the characteristic grid-group combinations of 4 ways of life. Most of those who employ the cultural theory approach operate with some version of this two-dimensional model. However, in an attempt to incorporate the dynamic force for change in social life we propose to construct a third dimension—namely, power. The *power* dimension expresses the extent to which an actor is able to effectively manipulate others and thereby the society he/she lives in. It is a measure of the effectiveness or efficiency of social action.[6] Once again the dimension has both positive and negative extension, manipulating and manipulated. (See Figure 2.3.)

Proposing a power dimension could be thought to be either bold or naive. It is not undertaken lightly. The purpose is to integrate the dynamic; to endogenize change. The leading practitioners of the two-dimensional approach account for change in various ways, and employ the two-dimensional model in tracing change, but we maintain that they do not treat change endogenously. Commonly, change is put down to more or less major disruptions of social life, such as war, famine or plague, or to some kind of choice or accident of life-style, such as unexpected inheritance, or ambition. Basically, practitioners of the two-dimensional model maintain that change comes about when expectations

Figure 2.3. The Power Dimension

Powerless. Ineffectual. Manipulated.	Autonomy/ Passivity.	Powerful. Effective. Manipulating.
<<═══════════	0	═══════════>>
negative		positive

are not met, when the world turns out to be other than expected. Essentially, change is attributed to the element of surprise. But to be reduced to accounting for change with terms like *surprise* is less than ideal. Hence this attempt to endogenize change by means of introducing the third dimension—*power.*

Social control, a form of power, is fundamental to the grid-group framework. Explicit prescriptions and group pressures are an integral part of the grid and group dimensions, but the social control aspect of these is not explicitly operationalized. By adding the third dimension the various combinations of social regulation and group pressure can be used to explain the nature of power in various social environments.

The distinctions between this and other three-dimensional formulations must be noted. Hampton, for example, introduced a third dimension of *activity*—the extent of social interaction.[7] We doubt, however, that the quantity of interaction alone really measures anything interesting. There is an important qualitative element in action. By no means is all action equally effective. The *effective quantity* comprises both the quantity and quality of action, the amount and efficiency or effectivity of action.

The next step in developing a workable three-dimensional approach involves operationalizing these three dimensions. The grid and group dimensions have been operationalized in many previous studies, and there has been considerable progress in giving them metric scale.[8] Operationalization of the power dimension is another, more difficult problem. Power is notoriously problematic in the social sciences, and we do not suppose the operationalization of this third dimension to be a simple task.

Nevertheless, certain attempts at operationalizing (measuring) power can be noted, and might prove useful in the further specification of the three-dimensional schema. One such attempt is that of Harsanyi,[9] who extended Dahl's foundational definition of the dimensions of power by two, to number seven. He did this in order to establish a measure of power that encompassed both the opportunity cost of the exercise of power and the probability of its success. Such an approach takes up the Machiavellian notion of what power does, rather than the more mainstream Hobbesian notion of what power is.[10] The bias to individualism in Dahl's formulation notwithstanding, this is clearly an advance along the path toward measuring the efficiency or effectiveness of power. It is taken as evidence of the practical possibility of the exercise of operationalizing the power dimension in the grid-group-power schema.

While we do not underestimate the difficulty of operationalizing these dimensions, it is suggested that it is only worthwhile tackling these problems if the model can be shown to be useful at a general level as an heuristic device. Consequently we concentrate on establishing the *relative positions* of the key groups and/or players in grid-group-power space, and tracing the changes in those positions over time. This is more than adequate to the task of demonstrating the applicability and usefulness of the cultural theory approach in the case we explore.

Cultural theorists have identified five viable ways of life. Each viable way of life involves a specific combination of social relations and cultural bias. Positions on the three dimensions of grid, group and power are determined by the structure of an actor's social relations. By examining the structure of social organization or pattern of interpersonal relations an actor experiences in everyday life, one can position that actor and his/her fellows on the grid, group and power axes. This identifies the actors with one of the five characteristic ways of life. Let us look at the five characteristic ways of life.

The Hermit: For a few there is the central position on the three dimensions wherein the individual is neither regulated or rule following nor regulating or rule setting. Such an individual is in a position of autonomy from groups, experiencing neither inclusion nor exclusion, and in a position of passivity as regards power, neither seeking to exercise control over others nor having to submit to control by others. "This is the way of life of the *hermit* who escapes social control by refusing to control others or be controlled by others."[11] This is the zero grid, zero group and zero power position.

The hermit social type has been identified with the Sherpa-Buddhist community in the Khumbu region of Nepal.[12] The easygoing tolerance of the Sherpas is one of the most immediate and enduring impressions of those that have visited the Khumbu. Sherpa-Buddhist villages and homes are open to strangers, and the Sherpa is a free agent, not subject to prescriptions or obligations. The area lends itself to occupations in trade and tourism, which often take place outside the local community, and imply neither dominance nor subordination. Such apparent social foibles as avoiding mentioning the names of the dead equip Sherpas with a cut-off mechanism that separates them from their own history and makes autonomy sustainable.[13]

The key characteristic of the hermit is autonomy, and there are many examples of autonomous ways of life closer to home. Occupations such as truck driver, caretaker of a small office or marginal farmer provide niche environments in which the individual can engage in a relaxed and unbeholden self-sufficiency.[14] They are individual occupations that avoid group involvement, regulated or regulating activity and offer little scope for gaining or exercising power over others.

The Fatalist: Many people find themselves subject to social prescriptions regarding their behavior and relations with others, and are alone and excluded from key groups. Lacking the means of control over others, they tend to be subject to control by others. They are isolated, manipulated rule followers. Their sphere of individual autonomy is restricted, they have little choice about how they spend their time, with whom they associate, what they wear and eat, where they live and work, etc.[15]

This is the social environment of the nineteenth-century mill worker, or the modern-day non-unionized factory worker. It is low group, low power, and high grid. Un-unionized and atomized as a unit of labor, such a worker depends on his/her employer, while the employer is independent of the particular worker. The routines, rhythms and practices of life are prescribed by others; daily life is surrounded by regulations that are beyond the worker's control. Life is like a lottery for such persons, and such people are, for as long as their perceptions match their way of life, powerless to change it. Good times and bad come to such persons seemingly regardless of their individual efforts. Work as hard as they may, the bankruptcy of the factory owner or a down-turn in the world economy would put them out of work. Personal survival in the face of helplessness is the basis of this social environment. It is the environment of the fatalist.

The Individualist: Others are more in control of their situation, and yet are also individuated. Such people are neither bounded by group incorporation nor by prescribed roles. Far from being prescribed or proscribed in their actions these individuals are able to negotiate their relationships with others, and can use this freedom to attain power over others. They enjoy both self-regulation and the regulation of others. Free from the need to bow to the needs of the group, these individuals are able to operate individually and to exercise social control over their own destiny and over the destiny of others. Being relatively free from control by others does not mean that these individuals do not engage in exerting control over others. On the contrary, the individualists' success is often measured by the size of the following the person can command.[16] However, this is not the kind of following that the charismatic leader commands; it is control and influence over the atomized individuals or amongst a network of peers.

This is the low grid, low group and high power social environment of the self-made entrepreneur. Such a person rises by exercising a rugged individualism, and through pursuing the manipulation of others. Opponent of regulating prescription, and of rules in general, such a person will champion the free operation of the market as an impartial arbiter of worth and of reward. Attempts to regulate are shunned, and attempts to impose a hierarchy derided as unfair. Such a person prospers, but claims to serve everyone by employing enterprise in turning raw materials into the valuable commodities that everyone needs. Those that labor for such persons are regarded by them as participating in the wealth that they generate.

The Hierarchist: Other individuals find themselves in a social environment in which their lives are intimately involved in groups. They are highly regulated, and yet at the same time they are powerful and able to exercise social control over others. Subject to both the influence of the group to which they belong and with which they identify, and to a set of prescriptions, conventions and mores that circumscribe and define their social role and their relations with others, such individuals see themselves as the occupants of clear-cut positions in an hierarchical set of social relations. Despite the behavioral circumscription of "office" such individuals often enjoy considerable power over others. Such persons are regulated by conventions and rules, are members of a group with which they identify, on which they depend and to which they defer, and yet commonly seek to make use of the social control that their position offers.

This is the high grid, high group and high power social environment in which that a high-caste Hindu Indian villager or a young subaltern find themselves. The high-caste Hindu (Brahmin) in an Indian village is highly group conscious, and in a context in which the demarcation between people (castes) is at its most extreme. Interaction between those in-group, the village community, and those outside the group is limited. Notions of good and evil, purity and pollution, safety and danger permeate the structure of social relations. Moreover these relations have become solidified into a strict structure of observation and a hierarchy of positions. The Brahmin villager is regulated in all the detailed activities of his/her daily life. Every act, public and private, must be performed according to ritual observance, every activity undertaken only in regard to that observance. What one does, to whom one speaks, what and with whom one eats, whom one touches and how one behaves towards others in the family, the village and beyond are all prescribed by convention, and inherited (not achieved) by virtue of caste position. And yet by virtue of membership in a bounded group such a person enjoys considerable rights to land, water, resources and rewards, and, of course, the deference of fellow villagers. The prescriptions that impose such a heavy behavioral gridiron are not the consequence of being manipulated by others, but rather the means by which the individual collectively manipulates others.[17]

The Hindu villager is a distant and somewhat untypical example, but the subaltern in an army regiment is in a similar social environment:

> His time, his dress, his social relations, his recreations, even his eating, his drinking and his sleeping. . . . are almost totally imposed by virtue of his fairly lowly position within a complex hierarchical organization. He has to wear a well-cut suit on an informal evening in the mess and he has to wear expensive mess dress on a dinner night. All sorts of compulsory items, . . . are added to his mess bill.[18]

All the prescriptions place and confirm the young man as a member of a complex hierarchy and subsume his personal identity within that of the regiment. He becomes a rank or office rather than a person.

The Egalitarian: Other individuals are in a social environment wherein they experience strong group boundaries, but very little prescription. Such individuals are involved almost exclusively with people in the same group; their work and leisure, family and friends are all enmeshed in the close-knit community. Often such communities are unified by a cause for which all members strive and to which all are committed. Within the group or community there is little or no role differentiation; all are equals. There is little or no social control exercised within the group because all are committed. The major, indeed often the only sanction on members' behavior is expulsion from the group. This has two major implications. First, such groups commonly experience a considerable degree of factionalism and factional disintegration. Second, such groups engender a strong sense of boundary in an effort to discourage members from leaving. Often this takes the basic form of notions of good and bad—good inside and bad outside—of purity and of danger. With strong group identification and little role differentiation within the group such groups are egalitarian.

This is the high group, low grid and low power social environment of the egalitarian. Exercising little social control within the group, and prone to factional differentiation whenever a leader threatens to emerge and/or is called for, such a social environment is low power. Individuals in such a social position might include those who are members of a commune, a minority religious sect, political party or community-based organization.

Sections of the environmental movement have been identified with this social environment, as have left-wing political parties.[19] In Australia many members of the Communist Party of Australia could have been placed in such an environment. In our more remote Indian village it is possible that the lower-caste Hindu (Vaisya) villager could also be in this position. Highly group-conscious, he/she is powerless relative to the higher-caste person, and yet is not subject to the extreme level of social prescription and proscription. Vaisyas are usually merchants or

Figure 2.4. The Five Regions in Three-Dimensional Space

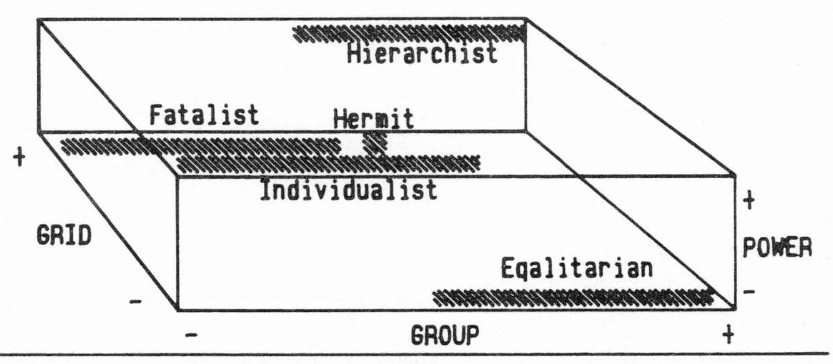

Source: M. Thompson. 1982. 'A Three-Dimensional Model.' In Mary Douglas. Ed.. *Essays in The Sociology of Perception* (London: Routledge & Kegan Paul). p.46.

artisans; they rely strongly on their caste group fellows in business and in life, and are dismissive of caste obligations, preferring instead the contractual form of association suited to their business lives.[20]

These five ways of life can be represented as positions on the three social dimensions of grid, group and power by the shaded areas of the cuboid. (See Figure 2.4.)[21] They are: the *hermit,* zero group, zero grid and zero power; the *fatalist,* low group, high grid and low power; the *individualist,* low group, low grid and high power; the *hierarchist,* high group, high grid and high power; and the *egalitarian,* high group, low grid and low power.

These five ways of life are made up of matched sets of social relations and cultural bias. The sets cannot be mixed in any other ways for long. A change in an individual's basic convictions about life, such as the way that he/she perceives human nature, necessarily changes the range of social action that he/she can justify engaging in to him or herself and to others.[22] Such a change in perception must change the range of activities and types of actions that it makes sense for the individual to undertake. For example, were someone who formerly believed that human nature was basically good, but inclined to be corrupted by social institutions, to come to see humans as unchangeably selfish, action aimed at reforming institutions in pursuit of the perfection of man/woman could no longer be justified.

Conversely, since no necessary causal direction is to be inferred, individuals who find themselves making a certain way of life for themselves must negotiate a set of values and beliefs that support and make sense of that way of life. One cannot for long actively participate in a way of life that does not make sense to oneself, and/or which one cannot justify to oneself and to others.[23] An egalitarian drafted into the army as an officer would soon find that he/she had to give up treating everyone, whatever his/her rank, as equals, or suffer great hardship until able to terminate his/her military service.

Taken individually, each of the five ways of life is a coherent and stable social environment, and the sets of social relations and cultural biases cannot be mixed in any other way. Moreover, each way of life depends to some greater or lesser extent on the existence of one or more of the others. Ways of life recruit from, leak to, and often define themselves in terms of, or in contrast to, one or more of the other ways of life. Were individualists able to rid the world of hierarchists there would be no extra market authority to enforce the laws of contract, and that would lead to the breakdown of the market and of the individualist way of life. If egalitarians eliminated or absorbed hierarchists and individualists there would be no target for them to be against, no cause to fight for and no cry to rally to. Thus robbed of their *raison d'être* the egalitarian way of life would disintegrate.[24] Without an atomized mass of fatalists and some hermits, egalitarians would have no one to convert and to fight for (on behalf of), individualists would lack human resources and consumers, and hierarchists would lack the masses that they seek to regulate. Each of the five ways of life requires that one or more of the others exists for its own coherence and viability.

Finally it is important to note that the five ways of life exhaust the possible stable social environments. Other combinations of grid, group and power are either not possible, or are not stable and enduring. We will take the four unoccupied regions of the cuboid social map in Figure 2.4. in turn, and examine the implications of these combinations of grid, group and power. Each will prove to be untenable.

A low grid, low group and low power social environment is incoherent because it is not possible to be at once rule-setting and powerless. If the rules that such an individual sets are followed by others then power is being exercised over the followers. If they are not followed, then the individual is not really regulating, and could not long make sense of a life spent making rules that no one followed.

A high grid, low group and high power social environment is also incoherent. An individual in such a position would be highly regulated (rule-taking), excluded from groups and thus not organized, and yet at the same time be powerful. Such a combination is unimaginable. An individuated, rule-taker cannot be powerful.

A high grid, high group and low power social environment is untenable. An individual in such a position would be assigned a clear-cut place in the hierarchy, and would be strongly committed to the group. Both imply an exercise of social control. The group depends on a boundary definition and maintenance that necessarily involves a manipulative differential treatment of those within the group and those outside. Likewise, a hierarchy is maintained in its exercise. To take one's place is to differentially treat those above and those below, and, more subtly, to differentially treat all the gradations between. To hold office is to exercise that office, which necessarily involves the exercise of control over others—the exercise of power.

Similarly, a low grid, high group and high power social environment is unsustainable because to be rule-setting and manipulative in the exercise of power must undermine the sense of group or community, and tear the group apart. Group coherence depends on a good inside/evil outside perception. Maintaining the clarity of the group boundary depends on maintaining a clear distinction between good and evil. This generates an extreme, polarized conceptualization, and mitigates against any perception of a continuum between, or gradations of good and evil. Hence all within group is extremely/equally good and all outside is extremely/equally evil. Manipulative control and egalitarianism are not compatible, because the manipulative control of equals cannot be justified. Egalitarians seek to convert not control, to persuade not prescribe.

To date we have identified the five ways of life largely in terms of the social relations involved. This is only half the story. The other key element of each way of life is its cultural bias, that is, the shared values and beliefs, underlying orientations or ideology that make coherent and sustain the particular way of life. "Operating under the Durkheimian hypothesis that the structure of symbolism parallels the structure of social life, we would expect symbolic systems to vary . . . in much the same way that social environments do."[25] Hence, each set of social relations has a matching way of thinking. We will now turn our attention to mapping these five ways of thinking, styles of thought or cultural biases, and matching them to their associated social relations.

The group dimension measures an individual's level of identity with a group, so one would expect it to also reflect the degree to which the individual sees himself/herself and his/her society as part of a natural universe.[26] One can expect people in a high group position to see society and nature as an integrated system, to be defensive and fearful of outside threats. Those in a low group position are inclined to see society and nature as separate, and to embrace new ideas in an open opportunistic way. The contrast in the goal of symbolic action between high and low group is that between preservation and continuity of the group and ego-centered self-preservation.

People in a high grid position could be expected to see society as routinized, and to be the defenders of a system of asymmetrical relations, of distinctions, of rules and conventions of behavior intended to distinguish between various social levels. Those in a low grid position are inclined to cross-cut any hierarchical or imposed rigidity of social position, and to be more spontaneous, flexible and personalized in both their lives and symbolic representations.

The underlying assumptions most basic to styles of thought are those concerning the nature of the world and of fellow human beings. Looking at orientations to the natural world ecologists have identified a few characteristic styles for the management of ecosystems. Confronted by the same situations, managing institutions respond differently, and yet in a patterned way.[27] These different responses are grounded in different views of the nature of nature.

The individualist way of life generates a global equilibrium conception of a benign nature. The strategy of a managing institution in an individualist social environment is *laissez-faire.* Nature is seen as a resource, trial and error are justified, and exploitation permitted. If everyone does his/her individual thing an invisible hand will ensure an optimal outcome, and equilibrium will be restored. Hence unfettered individual exploitation is justified.[28]

The egalitarian way of life generates the opposite notion of nature being ephemeral. The merest change can trigger a complete collapse of the system. Sanctions to prevent the collapse of the system must be observed. Ideally, people retreat to live in small communities, in harmony with nature and each other. The management strategy is to treat the ecosystem with great care. There is, in other words, justification for a communal, de-centralized life style that respects the fragility of nature.

The hierarchist way of life generates a perception that nature is at once perverse and tolerant—that is, forgiving most events, but vulnerable to shocks that might overpower the equilibriating mechanism. The hierarchist ecosystem management strategy is to regulate against unusual or shock occurrences. The goal is to ensure that the exploitation of nature does not become too exuberant, but rather stays within reasonable bounds. The creation and maintenance of the bounds of reasonableness become the province of experts, and all activity has to be managed and monitored. There is, in short, a justification for bureaucracy—a carefully managed hierarchy.

The fatalist concept of nature, by contrast, is that it is capricious. The fatalist management strategy, if such it could be called, is simply to cope with events as and when they occur. It is completely reactive. No one knows what will happen; there is no justification for acting or for active management. Life is a lottery. Once again this is a justification for the life style of fatalists; it promotes a reactive, do-nothing stance.

The hermit's life style generates a fundamentally different view, in which it is not a case of man acting on nature, but of an integrated system in which man and nature are one. It makes no sense to study nature and to then act on it in a strategic fashion in the light of that study, because for the hermit the first action, if not the act of study, changes the whole system. One is reduced to contemplation since the system is unknowable and the available strategies mutually contradictory. Practical management is impossible, contemplation and reverence the only sensible response. And, of course, this justifies the non-participative hermit life style.

The pattern as regards perceptions of human nature is similar. Particular perceptions of human nature are persuasive to people who are involved in particular structures of social relations. This is not an exploration of what human nature is like, but rather an exploration of the patterned nature of the various available conceptions of human nature. The way a person behaves toward others depends on that person's conception of human nature. If human nature were otherwise, that particular way of acting could not be justified.

This is of vital importance for the social sciences. A great deal of the most basic differences between the major approaches to a given discipline hinge on the largely taken-for-granted conceptions of human nature that underpin them. Compare, for example, the basic notion underlying Adam Smith's invisible hand and self-seeking humans with Karl Marx's notions of class and false consciousness. They are approaches built on fundamentally different conceptions of the nature of human nature.

Each of the five ways of life we have identified has a distinct conception of human nature. For the individualist, human nature is stable and enduring, and pervasively the same; humans are self-seeking. Unchanging and known, the way to organize society is to harness selfishness, rather than to try to change it. The natural system is that which brings into harmony individual selfishness and public benefit. Such a view justifies unbridled competition, and gives free rein to individual enterprise.[29]

Egalitarians conceive of human nature as basically good, but often corrupted by institutional manipulation. Importantly this establishes that human nature is malleable; if humans can be made bad, they can also be made good. Uncooperative, selfish behavior is seen as the product of evil institutions, and thought to lead to false consciousness.

Hence, there is justification for a non-coercive (low grid) and cooperative (high group) social organization—the egalitarian way of life.

The hierarchist view of human nature is a more complex conception of a balance or battle between the higher spiritual/intellectual man/woman and the baser animal man/woman. Born into the baser animal realm, hierarchists believe that the aim of life is to civilize the human. Hence there need to be very clearly defined stages of human development in which the various more-or-less civilized members of society are placed, and the development of the individual needs to be managed according to strictly controlled criteria. In short, a hierarchy is the necessary form of social organization.

To the fatalist, human nature is unpredictable. One never knows what to expect; people can be benevolent one minute and hostile the next. This cannot be predicted, changed or understood. The only plausible response is a fatalistic acceptance of the lottery of life.

For the hermit each of these conceptions is partially true, but it is more important that no individual has the right to manipulate another. Hence, good and bad people are at base simply what they are, and are as they are meant to be. Thus withdrawal and passive contemplation are the only sensible courses of action—the only courses left open.

What is important about these brief sketches is that they shift the basic questions from the great imponderables of what are nature and human nature really like to the question: How are perceptions of nature and human nature socially constructed? The aim becomes to explain why people see things the way they do, how and why different people in different social environments see things differently, what the patterns to these ways of seeing things are, and how this helps us both to understand choices and to make better choices.

This is an extension rather than a replacement of rational choice theory. It suggests that there are up to five equally rational choices available, because there are five possible ways of life each with its coherent sets of social relations and cultural biases. Rather than endlessly pursuing the question of which is rational, this approach examines the contextual rationality of each and asks which is the most appropriate under the circumstances.

The hermit is withdrawn from society. The hermit's environment promotes a detached attitude, wherein society and nature are complex collections of individual units. Systemic unity is little more than complexity and the awe that this generates. The fatalist does not generate an elaborate symbolic system; it is imposed by others. Unable to generate a strategy for personal success in this world or for salvation in the next, the fatalist is reduced to a rather mechanical participation in ritual. The individualist, by contrast, separating the individual and society and the individual and nature, seeks mastery over both. Such a person must take an active role to take advantage of society and of nature. Personal performance and personal (re)interpretation are characteristic. This is an orientation that leads to the type of critical review and reinterpretation that has been fundamental to the development of modern science, politics and religion. The hierarchist seeks to maintain the balance between society and nature, and to support the status quo. Often this requires an almost infinite elaboration, and an involution that symbolizes the intrinsic complexity of things in form and in ritual. Traditional doctrines are embellished and refined; they are not critically reviewed. The egalitarian sees society and nature linked in disequilibrium. Both are fragmented into antagonistic groups. It is a predatory world. Membership in a group affords protection so long as the group boundaries can be maintained. This involves personal divination, and the expulsion of evil influences. There is no single enduring ritual or symbolic structure.[30]

So, following and expanding on the examples of previous studies[31] we can match the five sustainable socio-cultural positions or ways of life we have identified with the following summary sketches of styles of thought or cultural biases. The *hermit* is inclined towards mysticism and pluralism, and is uncommitted. The *fatalist* is inclined towards eclecticism, revels in intrinsic complexity, adopts an empirical method, seeks order through routine, is open to external influence, prefers ego-oriented action, and has an unelaborated symbolic system. The *individualist* is inclined towards ultimate simplicity, adopts an abstract method, seeks goals of cause, is open to external influence, prefers ego-oriented action, and has an elaborated but simple symbolic system. The *hierarchist* is inclined towards ritualism, revels in intrinsic complexity, adopts an empirical method, seeks goals of order, is closed to external influence, prefers group-oriented action, and has an elaborated and elaborate symbolic system. The *egalitarian* is inclined towards fundamentalism/millenarianism, seeks ultimate simplicity, adopts an abstract method, seeks goals of cause, is closed to external influence, prefers group-oriented action, and has an unelaborated symbolic system.

The term *style of thought* has a particular usage in this context. Style is intended to convey a notion of change, even fashion. Style of thought is intended to convey a notion of a set of ideas, rather than that of ideas as individual things. These sets of ideas are grouped in part in respect of coherence, and in part in respect of commitment. They are effectively what would be called ideology, in a non-pejorative sense, by sociologists. What lies behind these styles of thought is a fundamental conception of the nature of nature and of human nature. These underpin the taken-for-granted, self-evident notions that provide

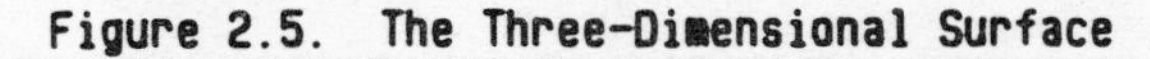

Figure 2.5. The Three-Dimensional Surface

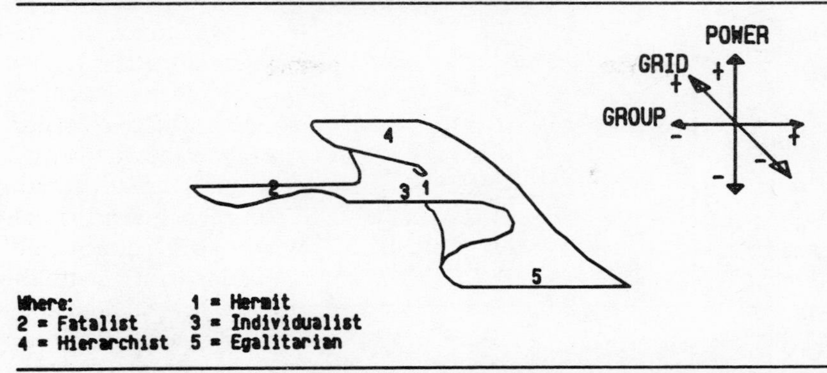

Source: M. Thompson, 1982, 'A Three-Dimensional Model,' In Mary Douglas, Ed., *Essays in The Sociology of Perception* (London: Routledge & Kegan Paul), p.50.

the foundation for a specific ideology, style of thought, heuristic or cultural bias—the cultural bias that sustains and is sustained by the structure of social relations or social organization experienced.

To get an idea of the space in which change is possible the five extreme positions that match the five possible ways of life can be joined to form a single continuous three-dimensional surface. Having established that only these five ways of life are viable, and that other combinations of grid, group and power yield incoherent and/or unstable ways of life that are either impossible or not enduring, it is no surprise that much of the space defined by the grid, group and power dimensions is uninhabited and uninhabitable. The only shape that links the five regions that represent viable, sustainable ways of life as a continuous surface is the rucked carpet shape shown in Figure 2.5.

Separating the ways of life in three-dimensional space, and then joining them through that space in this manner is immediately suggestive of the nature of change and of transition between positions. The relative extension of transitional surface suggests the extent of the transition, and the relative slope of the surface indicates the relative stability of the transitional positions. Flat regions suggest equilibrium states wherein a stable way of life emerges from the matching combination of a set of social relations and a cultural bias. Steeply sloping regions suggest disequilibrium states wherein the combination of social relations and cultural bias is unstable, and a viable way of life cannot be sustained.[32] The unoccupied regions of the cuboid, where the con-

Figure 2.6. The Fold (Cusp) Region

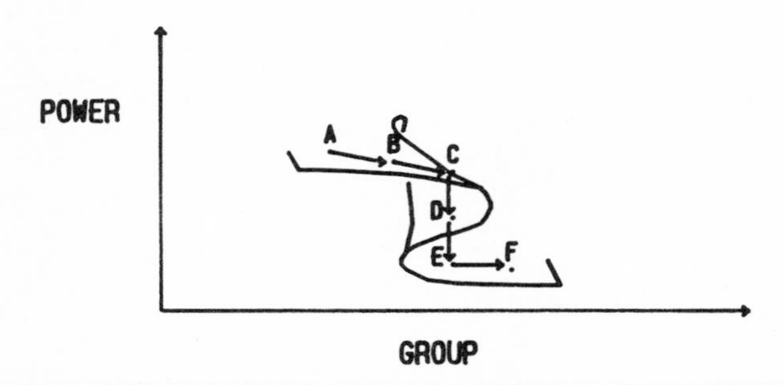

Source: M. Thompson, 1982, 'A Three-Dimensional Model,' In Mary Douglas, Ed., *Essays in The Sociology of Perception* (London: Routledge & Kegan Paul), p.49.

tinuous surface does not reach, are, as we have shown, incoherent or impossible ways of life.

In this construction something of the nature of both gradual and transformational change is revealed. Employing concepts from topology[33]—namely, cusp catastrophe theory[34]—to describe the implications of the derived shape, Thompson pointed to a key feature of the rucks or folds.[35] Put simply, the implication is that in the crucial fold regions gradual change along one dimension can lead to discontinuous change on another dimension, whereas, away from the folds, change is continuous on all three dimensions. By concentrating on a two-dimensional view of one fold area of the shape in isolation we can represent this pictorially. (See Figure 2.6.)

At points A and B there are single corresponding values on the vertical power dimension and on the horizontal group dimension. In the case of this diagram a relatively large movement along the group dimension produces a small decline in power. Importantly, at point C along the group dimension, within the fold region, there are suddenly three possible values for power for a given value of group. Two of these values, C and E, are stable upper-surface values, while D, on the inverted middle surface, is unstable.[36] There are, effectively, two alternatives available within this region, two equally viable alternative ways of life, each matched to an alternative combination of the key dimensions of social organization. With further movement along the group dimension, E to F in Figure 2.6., there emerges, once again, a single value correspondence on each dimension.[37]

There are, then, bimodal positions in the fold regions that are positions of discontinuous change and of choice, whereas away from the folds the situation is one of continuous, more or less gradual change. This suggests that in certain social environments, given certain specific characteristic life experiences set in characteristic forms of social organization, there will be a single unequivocal way of seeing and/or selecting, whereas, in other environments, there may be multiple alternatives possible. The former is a situation of relative stability and certainty with gradual change, while the latter is a situation of uncertainty, choice and discontinuous or revolutionary change.

The analogy in this construction with the well-known Kuhnian model of normal science and paradigm change is obvious.[38] Unlike Kuhnian formulations, however, we have here a potential explication of the nature of change, both gradual *and* discontinuous, and a means by which to relate this to social environment. This three-dimensional construction is a potentially powerful tool, which permits us to examine the inter-relationship between knowledge and action by highlighting the mutual determination of social environment (lived experience in participation) and cultural bias (committed world view or filter). Moreover, and most importantly, it permits the endogenous treatment of both continuous and discontinuous change. The three-dimensional schema sheds light on both the existential determination of knowledge *and* the nature of ideational/social change.

Conventional approaches have often operated in terms of a single dimension that is seen as either an evolutionary continuum or a dichotomy. The Weberian version, for example, involved a transition from the traditional to the rational, a process of rationalization, that culminated in a bureaucratic system. Other examples of such a framework are Durkheim's mechanical versus organic solidarity, and Tonnies's Gemeinschaft versus Gesellschaft.[39] The tendency has been towards a bifurcation that is at base individualism versus collectivism or agency versus structure. The debates that have arisen from these bases are unresolvable, and the schemas built from them lacking in sufficient variety to describe social life.[40]

The exploration of symbolic action and of symbolic systems must be founded at the level that symbolizes. As Ostrander has pointed out, this means the elimination of society as the unit of analysis, because societies do not symbolize—people do.[41] Hence it is necessary to focus on the social environments of individuals, and to recognize that there may be multiple social environments in society at any given time. An understanding of the inter-relation of action and perception, or the construction of a symbolic order, depends on an investigation of the spheres of action that are constrained by the social order. These include whom one interacts with, how one interacts with him/her,[42] and one's level of social control in those interactions—that is group, grid and power.

In summary, the model suggested herein is a development in what has become known as cultural theory. Cultural theory is about the cultural, that is, the organized and boundedly contextual settings' determining influence on ideas (as ideational sets) via the processes whereby micro social experience influence the development and maintenance (or change) of attitudes and values. It does not seek to link ideas to empirical events or to psychological phenomena, such as interests, nor is it the study of discourse *qua* discourse. It is, rather, the study of the inter-determination of social relations and perceptual schemas.

The task of the rest of this book is to address a case study, and to assess our model in terms of the light shed on actual historical events. We will now, therefore, turn to the story of the development of monetary theory and policy in England between 1797 and 1821. This exploration of the, so called, bullion controversy allows us to highlight the interpretive value of our model in relation to that of other approaches to the history of science and/or history of thought.

NOTES

1. Mary Douglas, 1966, *Purity and Danger: An Analysis of the Concepts of Pollution and Taboo* (London: Routledge and Kegan Paul); 1973, *Natural Symbols; Explorations in Cosmology* (Harmondsworth: Penguin); 1975. *Implicit Meanings: Essays in Anthropology* (London: Routledge and Kegan Paul); 1978, "Cultural Bias," *Royal Anthropological Institute,* Occasional Paper No. 35; Ed., 1982a, *Essays in the Sociology of Perception* (London: Routledge and Kegan Paul); 1982b, *In the Active Voice* (London: Routledge and Kegan Paul); etc.

2. Douglas, Ed., 1982a, *Essays in the Sociology of Perception* (London: Routledge and Kegan Paul), p. 3.

3. M. Thompson, R. Ellis and A. Wildavsky, 1990, *Cultural Theory* (Boulder CO: Westview Press), pp. 5–6, citing Mary Douglas, 1982b, *In the Active Voice* (London: Routledge and Kegan Paul).

4. Ibid.

5. It is important to note the distinction between this formulation and that of some others. Borrowing a language of networks from anthropology Thompson equated grid with the actors' centrality and/or peripherality in networks. Not only is this not faithful to the original formulation by Mary Douglas, it also raises what seems to be a serious problem. Interpreting grid as network involvement leaves it insufficiently distinct from group (group involvement). There is a danger of grid and group measuring the same thing, and thus failing to be independent variables; failing to be two dimensions. It must, therefore,

be stressed that in this formulation grid is taken as a measure of the extent of regulation; a reflection of a regulative hierarchy. Refer to Michael Thompson, 1982a, "A Three-Dimensional Model," in Mary Douglas, Ed., *Essays in the Sociology of Perception* (London: Routledge and Kegan Paul), pp. 31–63. Although Thompson's interpretation on this is rather idiosyncratic, it does demonstrate that there are important variations within the formulations of key cultural theorists.

6. Action being oriented toward a goal; as distinct from mere behaviour—response to stimuli.

7. J. Hampton, 1982, "Giving the Grid/Group Dimensions an Operational Definition," in Mary Douglas, Ed., op. cit., pp. 64–82. Again the purpose of this example is simply to demonstrate that a variety of three-dimensional cultural theory models exist, and that there are important differences between them.

8. Refer to J. L. Gross and S. Rayner, 1985, *Measuring Culture; A Paradigm for the Analysis of Social Organization* (New York: Columbia University Press); Hampton, 1982, op. cit., Celia and David Bloor, 1982, "Twenty Industrial Scientists; A Preliminary Exercise," in Mary Douglas, Ed., op. cit., pp. 83–102; G. Gaskell and J. Hampton, 1982, "A Note on Styles in Accounting," in Mary Douglas, Ed., op. cit. pp. 102–112; G. Mars, 1982, *Cheats at Work; An Anthropology of Workplace Crime* (London: Unwin Paperbacks); B. Wynne, 1983, "Technology as Cultural Process," Working Paper, wp-83-118, *International Institute for Applied Systems Analysis* (Laxenburg, Austria); etc.

9. J. Harsanyi, 1971, "The Dimension and Measurement of Social Power," in K. W. Rothschild, Ed., *Power in Economics; Selected Readings* (Harmondsworth: Penguin), pp. 77–96.

10. See S. Clegg, 1989, *Frameworks of Power* (London: Sage), for an explanation of this distinction, and for an overview of the development of the notion of power in the social sciences.

11. Thompson, Ellis and Wildavsky, 1990, op. cit., p. 7.

12. M. Thompson, 1982b, "The Problem of Centre; An Autonomous Cosmology," in Mary Douglas, Ed., op. cit., pp. 302–327.

13. Thompson, 1982b, op. cit., p. 308.

14. Thompson, Ellis and Wildavsky, 1990, op. cit., p. 10.

15. Ibid., p. 7.

16. Ibid., p. 7, citing Mary Douglas, 1982b, op. cit.

17. Ibid., p. 9.

18. Ibid., p. 45.

19. See Wynne, 1983, op. cit., Mary Douglas and A. Wildavsky, 1982, *Risk and Culture* (London: University of California Press), and S. Rayner, 1979, "The Classification and Dynamics of Sectarian Organisations: Grid/Group Perspectives on the Far Left in Britain," Unpublished Doctoral Thesis, (London: University College).

20. D. Ostrander, 1982, "One- and Two-dimensional Models of the Distribution of Beliefs," in Mary Douglas, Ed., op. cit., pp. 11–13.

21. Closely following the development pioneered by Thompson, 1982a, op. cit., pp. 44–45.

22. Thompson, Ellis and Wildavsky, 1990, op. cit., p. 2.
23. Ibid., p. 2.
24. Ibid., p. 4.
25. Ostrander, 1982, op. cit., p. 25.
26. Ibid., p. 25.
27. Thompson, Ellis and Wildavsky, 1990, op. cit., p. 26.
28. This and the subsequent sketches borrow heavily from Thompson, Ellis and Wildavsky, 1990, op. cit., pp. 26–29.
29. This and the subsequent sketches borrow heavily from Thompson, Ellis and Wildavsky, 1990, op. cit., pp. 35–37.
30. These summaries borrow heavily from Ostrander, 1982, op. cit., pp. 26–29.
31. See, for example, M. Rudwick, 1982, "Cognitive Styles in Geology," in Mary Douglas, Ed., op. cit., pp. 219–41; B. Wynne, 1983, op. cit.; D. Bloor, 1982, "Polyhedra and The Abominations of Leviticus: Cognitive styles in mathematics," in Mary Douglas, Ed., op. cit. pp. 191–218; D. Ostrander, 1982, "One- and Two-dimensional Models of the Distribution of Beliefs," in Mary Douglas, Ed., op. cit., pp. 11–13, etc.
32. Thompson, 1982b, op. cit., p. 305.
33. For further reference on the way in which such a shape is derived, and regarding the properties of it, see R. Thom, 1975, *Structural Stability and Morphogenesis. An outline of a general theory of models,* (Reading, Mass: W.A. Benjamin), translated by R. Thom and D. H. Fowler. See also M. Thompson, 1979, *Rubbish Theory; The creation and destruction of value* (Oxford: Oxford University Press).
34. While problems associated with the application of Cusp Catastrophe Theory are acknowledged (refer, for example, to R. S. Zahler and H. J. Sussmann, 1977, "Claims and Accomplishments of Applied Catastrophe Theory," *Nature,* Vol 269, pp. 759–63) its value as an heuristic device is suggested.
35. Thompson, 1982a, op. cit., p. 50.
36. Ibid., p. 47.
37. Note too that the folds are both located in a region which approximates zero on the group dimension. Given that it has often been noted that people who discover or invent—people, that is, who suddenly see things differently—tend to be people on the fringe, people who are neither actively included nor excluded, the model would appear to be coherent. The model "fits the facts."
38. T. S. Kuhn, 1962, *The Structure of Scientific Revolutions* (Chicago: Chicago University Press).
39. Ostrander, 1982, op. cit.
40. Thompson, Ellis and Wildavsky, 1990, op. cit., p. 3.
41. Ostrander, 1982, op. cit., p. 14.
42. Ibid., p. 15.

CHAPTER 3

Bullionism and Irish Currency

*Which philosophy is to be the dominant one of a society is one of the chief objects of the social struggle within that society. Each group has its own interpretation of the world, and seeks to make it the universally accepted one. Thus, theoretical discussions may be conceived of as incidents in the general struggle for power.**

Each of the following chapters concerning the case study will follow the same basic format; each has four main parts. The first will be a review of the overall socio-economic and political setting of the (sub)period, undertaken with the view of recontextualizing the specific points of debate and controversy. Rather than aiming at a comprehensive coverage, the summaries will note the particular historical changes that highlight the differentiation of the experiences of the leading social groups involved in the bullion controversy. The second part will involve a brief cultural analysis of this setting in order to outline our expectations regarding the bullion controversy. The third part will take a more detailed look at the development of the monetary arguments put forward. Accepting the broad consensus over what were the main theoretical contributions in the period, we focus on links between the setting or context and the nature, mode and even content of the ideas and arguments. This study is not intended to be exegetical. The aim is to treat the arguments as mediate presentations of an underlying structure of social relations, and to correlate the social relations with their characteristic cultural bias. The fourth part of each chapter will involve a cultural analysis of the contextual and textual situation(s) outlined in parts one and three, together with some critical comparison with other more conventional historiographic approaches. In this way

K. Mannheim, 1953, "Essays in the Sociology of Knowledge," in P. Kecskemeti, Ed., *Essays on Sociology and Social Psychology, By Karl Mannheim* (London: Routledge and Kegan Paul), p. 25.

the contribution of the suggested approach can be assessed, and a comparison with other approaches made. This pattern of presentation is dictated by the underlying purpose of the case study, which is to test the method against the historical evidence.

In this chapter we will cover the period which saw the emergence of monetary theory in debate concerning Irish currency circa 1797 to 1803. Before we begin it is necessary to be familiar with the use of a few key terms that will appear throughout the study. *Restriction* refers to the Restriction Acts, and more generally, along with *suspension,* to the situation wherein the Banks of England and Ireland were allowed to cease to honor the promise printed on their notes to "pay the bearer on demand" in specie, that is, to exchange a paper pound note for a pound of metal—gold and/or silver. *Resumption* refers to ending restriction or suspension, that is, to resuming payments in specie.

THE SETTING: 1797–1802

The French Revolution signaled a great change in intellectual life in England. Essentially it acted to polarize modes of thought and political consciousness. The whole of the nineteenth-century can be said to have been characterized by the steady progress to victory of rationalism. It was a rationalism that arose in opposition to both Aristotelian scholasticism and the natural law philosophy of the Renaissance. It opposed the qualitative ideas of Aristotelianism *and* the magical/analogical thinking of natural law philosophy. Whereas in natural law thought, logical and historical origins were treated together, they were increasingly separated with the progress of rationalism—the logical toward bourgeois rationalism, and the historical to a form of conservatism which grew up in response, a kind of traditionalism become conscious and oriented. The contrast between Smith's *Nature and Causes of the Wealth of Nations* and Ricardo's *Principles of Political Economy*[1] testifies to this division. Ricardo deduced, while killing off the history that had been so central to *The Wealth of Nations.*[2]

The constitution was a major concern of the period. At that time the main axis of contention in regard to the constitution was that between unquestioning defenders and those who felt that it had been corrupted by abuses and patronage, and who sought, therefore, to restore it to its former purity. The aim of the reformers of this type was to reassert the post–Civil War independence of the House of Commons. They did not seek to wrest power from the other elements of the triumvirate: the monarchy and the House of Lords.

The 1781 surrender to American republicans precipitated a constitutional crisis in England, and led to the replacement of the predominantly Whig ministries with a basically Tory regime under Pitt in 1783.[3] Consequently any progress that the Whigs had been making towards wresting constitutional power from the King was halted, and indeed reversed. Pitt wanted to get rid of some of the worst executive abuses, but he did not aspire to claiming power over the monarch. He did not wish to change the constitutional order, and was ready to follow the King's *will.*[4]

During the revolutionary period Pitt and his ministry represented stability and order, an old form of order. In this political and intellectual climate men of *principle,* of reason, lost ground to the defenders of the old constitutional and social order(s). Thus the progress of rationalism was interrupted by events in France. Under these circumstances English politics in the era was a continuation of the eighteenth-century structure of unstable constellations of fragmented small groups and of outright independents, rather than of unified parties. These groups clustered around leading people, charismatic leaders; they were not forged from agreement over principles. Parliament was an arena of personal charisma and interests and of pragmatic administration; it was not yet one where party political identification according to principle played any significant role.

In the early days of the French Revolution the whole pace of political life in England was accelerated. There was widespread support for the overthrow of the backward-looking *Ancien Régime* amongst the leading figures of the intellectual community. Societies and clubs sprang up, and the ideas of the revolutionaries and their supporters were widely discussed. One of the most important of these clubs was the London Corresponding Society, which was formed under the leadership of the shoe maker Thomas Hardy in early 1792.[5] Many provincial societies followed the pattern of the London Corresponding Society over the next few years.

The political pace was, however, soon to be slowed again by harsh repression from the Pitt-led government. Fearful of the turmoil spreading from France, Pitt introduced such repressive acts as the Traitorous Correspondence Act in 1783, the suspension of Habeas Corpus in 1794, the Seditious Meetings Act of 1797, and the Act of Union with Ireland in 1800 in the face of the 1798 Irish rebellion.[6]

The wars with France (the revolutionary and Napoleonic wars 1793 to 1815) had three distinct phases. First was the war with revolutionary France, declared in 1793 when France was under the leadership of the revolutionary directorate. This phase ended with the Peace of Amiens in March 1802. Second was the war with Imperial France from 1803 to 1814; and third, the so-called Hundred Days from February 1815 to

Waterloo.[7] Each of these phases engendered different popular attitudes in England. Our concern at the moment is with the first phase only.

Whereas the war with Imperial France evoked a good deal of patriotic fervor and a rally in defense of King and Country in the face of French expansionism, the war with revolutionary France was often unpopular at home—though decreasingly so in later years. Many of liberal political and economic persuasion, the so-called Friends of Peace,[8] welcomed the changes in France, and so looked unfavorably upon the war. Consequently, in the late 1790s, leaders in England were polarized and divided over the prosecution of war on France.

The Pitt-led government enjoyed a considerable degree of economic success in its first ten years in office, 1783–93. During that period Pitt, a self-confessed disciple of Adam Smith, led Britain from the depths of a depressing defeat at the hands of America to a strong economic position at the start of the wars with France.[9] Fortune, however, did not favor Pitt's administration in wartime as it had in peacetime. From the height of strength and optimism in 1793 Britain had plunged to the brink of defeat by 1797.[10] Indeed, it was not until some time after Pitt's death in 1806 that the war showed any sign of turning around in Britain's favor.

Though Pitt was not widely perceived as a great economic leader he was a great political leader. After the defection of the Portland Whigs to the Pittite camp in 1794 the Whig party also had a very considerable leader in Fox.[11] The chief effect of the leadership of these two men was to unite the two polarized camps. Under their guidance intra-party rivalry was held in check, and played a secondary role to bipolar opposition. Effectively, these leaders personified and extended the polarization which the French Revolution and Revolutionary Wars produced in England.

Pitt's wartime policy set the tone for the whole of the wars period. It consisted in the reliance on a reasonably well equipped if rather dispersed Navy, and the use of gold to financially support Britain's European allies.[12] Despite wartime disruptions Britain's overseas trade *value* continued on trend to increase.[13] It was not really until 1807–08 that the war-related trade disruptions led to any great economic disruption. One of the major contributory factors to this apparent trade expansion during the revolutionary wars was that the aid which Britain gave to her allies was spent, in part, on British goods.[14] A note of caution must, however, be sounded here because of the interrelation of price changes and apparent trade value expansion. It is likely that, while trade value increased, trade volumes were little changed in the war years before 1807–08.

Table 3.1. The Schumpeter-Gilboy Prices Index 1793 to 1822 (1701=100)

Year	Index	Year	Index	Year	Index
1793	129	1803	156	1813	243
1794	136	1804	161	1814	209
1795	147	1805	187	1815	119
1796	154	1806	184	1816	172
1797	148	1807	186	1817	189
1798	148	1808	204	1818	194
1799	160	1809	212	1819	192
1800	212	1810	207	1820	139
1801	228	1811	206	1821	139
1802	174	1812	237	1822	125

Source: C. Cook and J. Stevenson, 1980, *British Historical Facts 1760 to 1830* (London: MacMillan Press), p. 182.

The problem that faced the English public during the first phase of war was a sharp increase in commodity prices from 1799 onwards. (See Tables 3.1 and 3.2.) High food prices meant high land rents and high tenant farmer profits, but they also meant hardship and starvation to the poor, even the working and, so-called, deserving poor. A considerable degree of disquiet and distress due to high prices emerged during this first phase of the wars, and played an important role in the early phase of the bullion controversy.

The most common popular response to the economic distress that the public felt was rioting and violent attacks on property. This was

Table 3.2. Wheat Prices Per Quarter, and Bread Prices Per Loaf

Years	Wheat Prices (Shillings per quarter)	Bread Prices (London) (Pence per 4 lb. loaf)
1750-54	31.25	5.1
1755-59	36.54	5.6
1760-64	32.95	4.9
1765-69	43.43	6.6
1770-74	50.21	6.8
1775-79	42.80	6.3
1780-84	47.32	6.7
1785-89	44.92	6.1
1790-94	49.57	6.6
1795-99	65.67	8.8
1800-04	84.85	11.7
1805-09	84.57	12.2

Source: P. Mathias, 1969, *The First Industrial Nation: An Economic History of Britain, 1700-1914* (London: Methuen and Company), p. 474.

both an urban and a rural phenomenon, involving food riots, machine breaking, rick burning, and more.[15] While riots were a common response to the economic distress more sophisticated observers began to link high prices with the wartime economic policy of the Tory government. Pitt used the money market as a means of financing the war through debt creation. The extent of the national debt expansion was unprecedented. From 228 million pounds in 1793 the debt blew out to 876 million pounds by 1815.[16] The interest charge on this debt alone in 1815 exceeded the whole of government outlay circa 1792. Despite the introduction of income tax, which Pitt hoped would obviate the need to take large government loans,[17] over 40 per cent of war-related expenditure between 1799 and 1816 was met by borrowing on permanent funded debt. Such wartime expenditure on credit was, of course, highly inflationary.

One of the major non-economic effects of such large-scale wartime funding was to raise public awareness of government finances. The government had to publicize its needs in order to raise the loans, and had, consequently, to justify its actions.[18] The revolutionary and Napoleonic wars brought economic analysis into popular discourse for the first time. Rising food prices, heavy taxation, rising population and poor relief demands, the fear of monetary disorder, and the extent of the burden of public debt became pressing public issues to be urgently addressed.[19] The wars came to be portrayed on both sides as a war of finance, and this meant that systems of war finance came under close scrutiny. The fact of war was controversial and divisive; the effects of the war were controversial and divisive; and, perhaps most importantly in our context, the financing of the wars was also controversial and divisive. It is important, therefore, to look closely at the system of war finance that was used.

War Finance

The Bank of England had been established in 1694 primarily as an agent of government finance.[20] It held a monopoly on joint stock banking in England. By the late eighteenth-century it had become the *de facto* central bank in an increasingly complex banking system. While the Bank of England held, and used, the right of issue, other London banks had, by convention, ceased to issue notes. Outside the metropolis, however, country banks continued to issue their own notes. These often formed the majority circulation within their locales. One crucial aspect of the system was that outside London the Bank of England and the country bank issues circulated together; they were not mutually exclusive.

The situation in Ireland was somewhat different. The Bank of Ireland did not have sole issue rights in either the country or in metropolitan Dublin. There were a number of other powerful private banks issuing notes in the metropolis. Consequently the Bank of Ireland had even less control over the money supply in Ireland than the Bank of England did in non-metropolitan England. In England London-based private banks held the greater part of their reserves in Bank of England notes. The country banks primarily held London private bank deposits and Exchequer and private bills that could be converted to Bank of England notes as required.[21]

It is also notable that there was not, technically at least, a free metallic system of exchange prior to the restriction of payments in specie at the Banks of England and Ireland. It was illegal both to export bullion from melted coin of the realm, and to melt full-weight coin of the realm.[22] Thus, while gold and silver were freely convertible to coin, coin was not freely convertible to gold and silver.[23] It was generally held that these restrictions were so widely disregarded as to have been practically ineffective under most circumstances, but whether they were ineffective under the unusual circumstances of war is a moot point.

The country banks in England enjoyed enormous growth in business. Their numbers were said to have doubled during the decades 1783 to 1793 and 1800 to 1810.[24] They were a new phenomenon, a new element in the system, and as such were totally unregulated save by their own basic commercial considerations. In times of crisis many failed. While it was widely held that the country banks provided an essential access to credit for industry and commerce outside the metropolitan area, insofar as they represented a new and destabilizing influence, they were often viewed with suspicion at that time.

The eighteenth-century practice for raising government funds was to issue loan stocks direct to the public. As time went by the size of the loans raised made it both increasingly difficult for the market to absorb such issues, and more desirable for the government to ensure that they were underwritten. Consequently a few businesses with large resources began to take on the role of loan contractor. Typically, these contractors compiled a list of subscribers, and took up part of the loan themselves. In this way they smoothed out the issues' market entry, and acted as *de facto* underwriter to such subscribers as might default. With the enormous growth in the size of government loans during the period of the war with revolutionary France this mode of contracting became established.

Once a loan contract had been settled for a given year the successful and the unsuccessful contractors had conflicting interests. The successful contractors wanted a rising market to enable them to sell their residual share of the funds at a profit, while the jobbers, being more often than not short of stock, sought to depress prices.[25] As the stakes increased, so too did the struggle between the two groups, between the successful and the unsuccessful contractors, become increasingly intense.[26]

The major government loan contractor in the 1790s was Boyd, Benfield and Company Bankers (led by Walter Boyd). The loans could be very lucrative for the contractors, and there were often suggestions that favoritism played a part in awarding the contracts. The 1795 loan, for which Boyd, Benfield and Company was the successful contractor, was a case in point. It turned out to be insufficient for the government's needs for that financial year, and it sought to raise another loan during its term. Boyd, Benfield and Company representatives complained that this was against the spirit of their contract and would inevitably depress the value of the loan stock they held. Pitt apparently agreed, and awarded the new loan to the same contractor without the usual competition and bidding. Other would-be contractors, including James Morgan of the Stock Exchange, complained that this was unfair.

The case generated considerable controversy, and ill feeling, and resulted in a House of Commons committee of enquiry,[27] which vindicated the ministry's decision. Though victorious in this battle, Boyd Benfield and Company lost the war. The contractor was denied the opportunity to contract for the loan of 1799, denied credit extension at the Bank of England, and subsequently failed in the commercial crisis of 1800–01.

One impact of the changing commercial situation in Britain, and of the increasing funding demands of government, was to draw forth a (re)organization of the London money market. Essentially the market had three centers: the Royal Mint, the Bank of England, and the Stock Exchange.[28] Up until the nineteenth-century Stock Exchange dealers were perceived as a low-status group. They were money grubbers, Jews, national and religious outcasts, and their occupation was perceived by everyone else as distasteful and evil; perhaps a necessary evil, but an evil nevertheless. The coffee houses of "Change Alley" were derided as the scenes of rude and clamorous behavior, and of underhanded dealing.

Legislation in the period reflected this attitude. Barnard's Act of 1734 was entitled *An Act to Prevent the Infamous Practice of Stock Jobbing.* Capturing the mood of the time it also set the scene for the developing organization of the Stock Exchange. Barnard's Act had the unforeseen consequence of placing jobbers beyond the law, such that they had to rely on their own self-regulation in practice.[29] It also acted to heighten the divisions of specialization between the various key market centers, and between the operative money market groups.

Those active in the money market at that time were typically drawn from one of three groups: large merchant houses, such as the Goldsmids; London bankers, such as Boyd, Benfield and Company, and Baring Brothers; and the brokers and jobbers who were specialists on the Stock Exchange, such as David Ricardo, James Morgan, John Steers, etc. The directors of the Bank of England were drawn from an inner circle of leading merchant houses[30] (the first group). Bankers did *not* generally become Bank of England directors.

At the turn of the nineteenth-century the Stock Exchange specialists were seeking to insinuate themselves (functionally) between the merchants and the bankers. They were seeking to raise the status of their trade by furthering specialization and interdependence in the money market. In short, the London money market consisted of three main competing groups, and, as the so-called Gentlemen of the Exchange pushed toward functional specialization, these groups became increasingly defined and distinct.

The increasing self-regulation played a major part in the moves to relocate and to reorganize the Stock Exchange circa 1801. Following the move to their own coffee house, called the Stock Exchange, in 1773 the regulation of the market was in the joint hands of two main committees: the Proprietors Committee and the Committee for General Purposes. In an effort to enforce some respectability and to raise the public estimation of the Exchange the two committees joined into one United Committee in 1801. They attempted to turn the Exchange into a subscription room, such that it would no longer be open to the public.[31] The "change-over did not go by any means smoothly; there were complaints from former users who failed to secure election, friction between proprietors and users,"[32] and disputes over the rules. The United Committee soon split up again.

A powerful subgroup was, meanwhile, going ahead with plans to move to new premises in Capel Court, and as the time of the completion of this new building neared its proprietors invited the proprietors of the old subscription room to join them. They refused flatly.[33] There followed a period of considerable animosity, during which the joining, resigning and realigning of participants with the various conflicting committees was rapid. The situation of the financial community in the City of London was emerging as one of differentiation and of increasing factional conflict. This was reaching a crescendo in the 1800–01 period.

Soon after war with France broke out in 1793 there was a severe commercial and financial crisis precipitated in large part by fear of invasion. During this short-lived monetary crisis of February to August

Table 3.3. Number of Bankruptcies Gazetted Over 29 Years to December 1811

Year	No.	Year	No.	Year	No.
1790	583	1797	866	1804	884
1791	612	1798	724	1805	958
1792	625	1799	557	1806	994
1793	1299	1800	736	1807	1067
1794	824	1801	884	1808	1101
1795	704	1802	947	1809	1110
1796	755	1803	920	1810	1792

Source: W. Cobbett, 1811, *Political Register* (London), Volumes 17, 18 and 20, p. 1.

1793 there was a sharp rise in the number of bankruptcies. These included banks and some of the larger commercial houses.[34] (See Table 3.3.) Though severe this crisis was quickly corrected by government initiative and intervention. It was a notable feature of this, and the subsequent crisis of 1797, that the Bank of England neither showed nor accepted any responsibility for either the control of money supply, or as lender of last resort.

There were many businesses which suffered during these crises, and many ruined or near-ruined businessmen who viewed the (in)actions of the Bank of England, pursuant wholly and solely of commercial self-interest, with dismay. As a result there developed a degree of animosity in business circles towards the Bank of England and its directors. It was under these circumstances that the question of the role and best organizational form for the Bank of England came onto the political agenda.

The collapse of banks during the crises of 1793 and 1797 produced the kind of crisis of confidence that led inevitably to a drain on the Bank of England specie reserves. By 1797 this was compounded with other war-related drains—the first being the expenses of war in the forms of both the deployment and maintenance of forces on the continent and at sea, and of loans (often more in the nature of gifts) to assist allies. Under the circumstances of trade disruption and general uncertainty these expenses were often met by remittances of bullion. Second, during 1793 and again in 1797 there coincided an internal drain on specie due to a run on English banks. This was provoked primarily by fears of imminent invasion. A drain to Ireland also occurred to cover a similar run within Ireland for the same reasons.[35] Third, there occurred a movement of specie to France following the collapse of the paper *assignats* in 1795, and the French return to a

Table 3.4. Bank of England Reserves (Million Pounds), 1795-1820

Year	Reserves	Year	Reserves	Year	Reserves
1795	5.4	1804	4.8	1813	2.6
1796	2.3	1805	6.7	1814	2.2
1797	3.1	1806	6.1	1815	3.0
1798	6.4	1807	6.5	1816	6.6
1799	7.0	1808	6.4	1817	10.8
1800	5.4	1809	4.0	1818	7.8
1801	4.5	1810	3.4	1819	3.9
1802	4.0	1811	3.3	1820	7.3
1803	3.6	1812	3.0		

Source: N. J. Silberling, 1924, "The Financial and Monetary Policy of Great Britain during the Napoleonic Wars: Part I - Financial Policy," *Quarterly Journal of Economics* (New York: John Wiley & Sons), Volume 38, p. 227. Copyright © John Wiley & Sons. Reprinted by permission of John Wiley & Sons.

metal standard. These events, entailing both internal and external drains, put considerable pressure on the Bank of England's reserves through the 1790s. (See Table 3.4.)

From the middle of 1795 there developed a sharp decline in the Bank's reserves. By August 1796 the Bank held only 400,000 pounds in coin and bullion while having liabilities of 16 million pounds.[36] Under these circumstances, and in the face of continuing high demands from the government for wartime credits, the Bank of England directors became alarmed. They confronted Prime Minister Pitt seeking help, and they received an Order of Council allowing the suspension of specie payments for notes on 27th February 1797.[37] This suspension subsequently became embodied in the Bank Restriction Act. The Bank of Ireland gained the right to suspend payments soon afterwards. The 1797 and 1798 Restriction Acts applied only to the Banks of England and Ireland.[38] Technically, all other banks were still obliged to redeem in specie. In fact, however, most did not do so.[39] Instead, they redeemed in Bank of England paper and London drafts, which constituted the greater part of their reserves. Thus restriction became the *de facto* practice.

Bullionists and Anti-Bullionists

It is possible, with a risk of gross oversimplification, to trace the following broad pattern of alignments as regards monetary theory and policy by way of introducing a more detailed analysis. This broad-brush review will allow us to begin to place the protagonists in their cultural context.

Anti-bullionists took the view that the special circumstances of high wartime expenditure adversely affected the balance of payments, that this could not be righted by trade because war had caused a disruption of trade and shipping, and that this was the cause of the continued exchange depreciation.[40] They commonly pointed to the fact that bullion itself varied in value, and asked why then should bullion alone be the standard by which to measure the value of currency.[41] In these circumstances they questioned the meaning that could be attached to the words *depreciation* and *excess issue.* They further argued that a resumption of payments would force the Bank of England, and thereby the country into insolvency, and thus prevent England and her allies continuing the war against France. Anti-bullionists were often Tories who placed defense of King and Country high on their agenda. They argued that restriction had the effect of breaking the links between events in Europe and England's money supply,[42] and they saw this as a valid wartime strategy. They suggested that Napoleon was quite cognizant of the power of credit as a military weapon,[43] and would not hesitate to attempt to destabilize Britain's financial position given the chance to do so.

Overlaid on these wider political aspects were more local ones of commercial power. The directors of the Bank of England held considerable commercial power and they were not keen to see it undermined, or to take the blame for events which they perceived to be beyond their control. They wished to guard their ability to undertake their business for profit, without facing demands for wider responsibility or undue controls. Under restriction their profit potential was very much greater since extension of credit was more a question of discretion than it would have been with convertibility. With the rapid growth in the number of country banks and the growing importance of the country areas in the emerging industrial economy there grew up a very powerful private banker lobby. It vied with the Bank of England on the issue of controls. The Bank of England pressed the government to control country and other private banks rather than the Bank of England, while the private bankers wanted to be left alone and argued for controls on the Bank of England.

Bullionists, on the other hand, argued that restriction had given the directors of the Bank of England *carte blanche* to issue paper money. Since it was clearly in the Bank's economic interest to issue as much paper as possible it was presupposed that it had done and/or would do so, and this was perceived to be the cause of the depreciation of bank paper against bullion, and of the pound against foreign exchanges.[44] It was their assumption that bullion, and only bullion, could be the standard of value of the currency that lent them the name bullionists.

That restriction applied to the Banks of England and Ireland alone was seen as a suspicious sign of collusion. The Bank of England was thought to have gained this privilege because it was the government's bank, because it would ruin the government if it recalled its loans. Often the bullionists were Whigs or radicals who were critical of the Tory government's war policy.[45] Many were not in favor of England prosecuting war on France, and opposed the wartime policy of the Tory ministry. The connection of the Bank of England with public finance and with serving the ends of war put it at odds with many bullionists. In economic terms bullionists were often liberals who opposed the Bank of England's monopoly position. Those amongst Bank stockholders were also critical of the Bank's poor profit performance.

The bullionists argued that, once the fear of invasion and the consequent drain of specie in internal runs had passed, restriction should have been lifted.[46] Even those who agreed that the 1797 alarm had been sufficient reason for restriction thought its continuation anachronistic. Indeed it has been said of the continuation of restriction for more than 22 years:

> Cash payments might have been resumed after a couple of months, and the Bank of England was quite willing. But no government involved in a great war is willing to give up so potent an engine for surreptitiously fleecing its subjects as an inconvertible currency . . . For "political reasons" the Restriction Act continued.[47]

We shall presently have reason to qualify this common assessment of the situation, but for now it does serve to highlight the one (macro) level of political reason. While the government and the Bank cooperated closely their power(s) were as one, but it is a notable point that the power lay with the Bank of England *not* the government. In that context it went against the grain, and contrary to the principles of the political reformers to allow such a vital economic weapon, such an instrument of power as money supply, to lie in hands entirely free from public accountability. They felt a keen desire for regulation and control over the Bank of England, both in order to enforce its responsibility as the central bank and to place the power over money supply in the hands of elected representatives.

Put simply, then, anti-bullionists were typically drawn from Tory ministerialists and/or Bank of England (merchant) ranks, while bullionists were drawn from amongst political liberals who opposed Tory war-making, and economic liberals who opposed the Bank of England's monopoly powers, and the power of the old merchant order.

AN ANALYTICAL PREVIEW

What does this brief outline of the political and socio-economic context tell us? In terms of our aim to relate social environment to style of thought, and thereby monetary theory, we might summarize it analytically as follows.

The French Revolution effected a polarization of the political leadership into liberal and Tory camps, and this polarization continued until late into the first phase of the war with France. With both Tory and Whig groupings having strong leaders this polarization would have been all the clearer. In the context of war finance the ruling Tory elite had to turn to the money lords, the emergent economic elite, for the financial means to prosecute war. This can be expected to have heightened their awareness of group inclusion and exclusion. The position of the Tory leadership was under siege politically and economically. In cultural theory terms this suggests that the leaders were in a high positive group position.

As regards the grid dimension it must be noted that those in the dominant Tory political leadership positions were the representatives of an old order, one based upon mutual obligations rather than market contractual relations. Theirs was a system of conventions, traditions and obligations. They were themselves highly regulated. Their life experience was of adherence to a strict code in all aspects of their lives—a position comparable to that of the high-caste Indian. Hence they must be placed in a high positive grid position, a regulated position. They were, of course, the ruling elite. Indeed the old-style Tory leadership was firmly in the box seat at that time, a high positive power position. (See Figure 3.1.)

Radical parliamentarians were much less enamored by, and involved in, the old order of things than the Tory elite. They can be placed in a lower grid position. With wartime repression of free speech and public meetings they were, however, clearly moving up grid towards a more regulated position. They can be placed in a relatively high group position since they were formed into an opposition that was characterized more by factionalism and groupings around charismatic leaders than by adherence to principles. The inclination of radicals was towards relative egalitarianism. They were also increasingly conscious of group differences under the polarizing influence of war. They were neither in power nor able, in the context of wartime repression, to increase their hold on political power. In the cultural theory position of egalitarians,

Figure 3.1. Position of the Tory Elite (1797-1802)

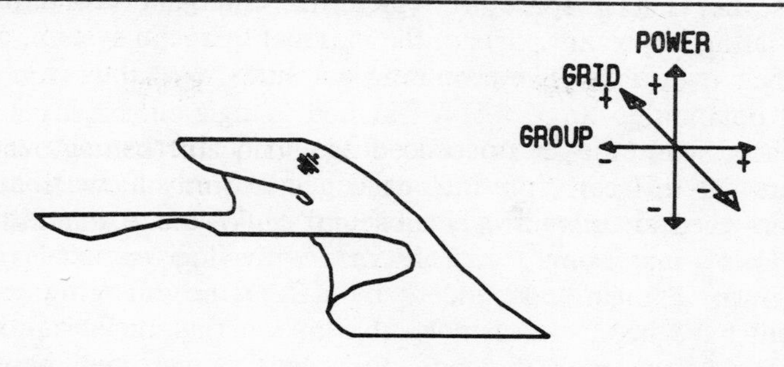

⁂ Indicates Position.

Source: M. Thompson, 1982, `A Three-Dimensional Model,' In Mary Douglas, Ed., *Essays in The Sociology of Perception* (London: Routledge & Kegan Paul), p.50.

then, Whigs and radicals can be placed relatively high grid and high power. (See Figure 3.2)

In an attempt to further their own exclusiveness the aristocratic elite tended to shun economic pursuits. Thus the rising economic elite were left to affect their own self-regulation in an increasingly important arena of national life. This aristocratic pursuit of exclusiveness amounted to

Figure 3.2. Position of the Political Radicals (1797-1802)

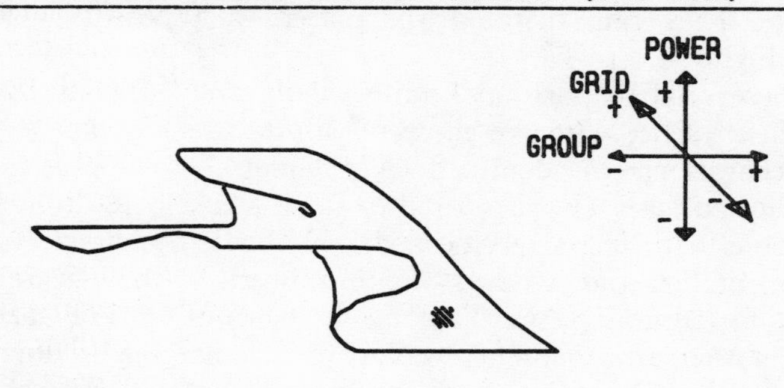

⁂ Indicates Position.

Source: M. Thompson, 1982, `A Three-Dimensional Model,' In Mary Douglas, Ed., *Essays in The Sociology of Perception* (London: Routledge & Kegan Paul), p.50.

sowing the seeds of their own destruction. In shunning economic pursuits they left a space in which the emergent economic elite were regulating important parts of the national financial system, and of course of their own lives. The economic parvenus were thus in a low negative grid position.

Faced with the aristocratic elite's exclusiveness the economic parvenus—at this early pre-industrial stage chiefly a financial parvenus—experienced an increasingly profound exclusion. An exclusion that was becoming increasingly out of kilter with their national economic importance. Though independent by nature the emerging economic elite would have been increasingly able to see that they were excluded. At the same time, with the economic needs of war, and the emergence of factional conflict in the London financial community, with functional differentiation and specialization in pursuit of status, the economic parvenus were moving up group. This rising economic class had not been organized into a group as such; they were merely excluded individuals. All the war-related changes were, however, acting toward the development of a group consciousness. Thus the parvenus should be placed in a high negative group position. A position wherein the individualistic were becoming conscious of the exclusion and were beginning to organize as a group.

The wars brought an increase in opportunities for the economic parvenus. These opportunities were, at this stage, mainly financial. The aristocratic ruling elite who so wished to prosecute war on revolutionary France had to turn to the parvenus for support, for the economic means of war. Hence the parvenus, who had almost free rein in the economic sphere, were becoming increasingly important in the national political sphere. They can, therefore, be placed in a positive power position. (See Figure 3.3.)

These relative positions, quite clearly and easily derived from historical evidence with the guide of cultural theory, can be equated with the characteristics identified in Chapter 2 as representative of the various cultures or ways of life. It is possible to link these relative positions with characteristic styles of thought, characteristic modes of argumentation and patterns of commitment to thought and value schemas—to cultural biases. Thus on the basis of cultural theory we could predict that any theoretical controversy emerging in this social environment would exhibit the following features:

1. The old Tory elite would argue a line characteristically hierarchist, that is, one impressed with intrinsic complexity and based on empirical methods. The underlying goal would be that of order—

Figure 3.3. Position of the Economic Parvenus (1797–1802)

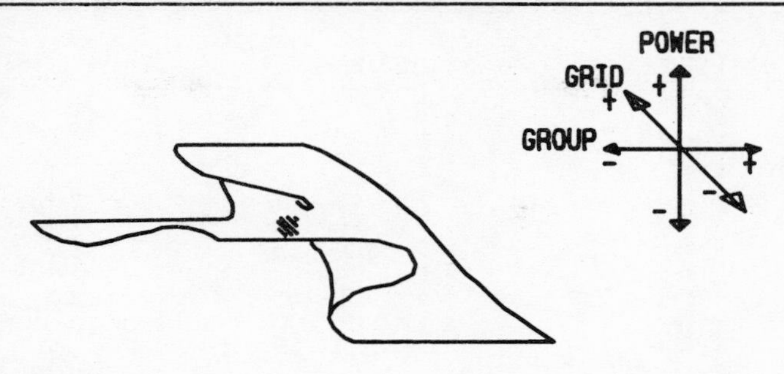

⁂ Indicates Position.

Source: M. Thompson, 1982, 'A Three-Dimensional Model.' In Mary Douglas, Ed., *Essays in The Sociology of Perception* (London: Routledge & Kegan Paul), p.50.

to maintain the order in which they lived and in which they ruled.

2. The emergent economic elite would argue a line characteristically individualist, that is, one impressed with ultimate simplicity, and based on abstract methods. The underlying goal would be the revelation of causal relations.
3. The Radicals would argue a line characteristically egalitarian, that is, one similar to the individualist, but more fundamentalist and with a less elaborated symbolic framework.
4. Given that it is the economic elite who are emergent, their style of thought can be expected to be the new one, the basis of the new theory, and the one that is the challenger to that of the Tory hierarchists.

We shall now turn our attention to the emergence of monetary debate, and see if these features, the characteristic styles of thought, that correspond to the styles of life and social environments we have identified, can be seen in the ensuing debate.

MONETARY DEBATE: 1797–1802

The initial passage of restriction was associated with fear of invasion. Through 1797 this fear was a very real one. By the end of 1797, however, immediate fears of invasion had passed. In this context the

Table 3.5. Percentage Deviations from Par, 1790-1808

Year	Spanish Dollars	Hamburg Exchange
1790	0.5	1.7
1791	1.6	2.0
1792	6.2	-2.0
1793	-0.5	4.6
1794	1.0	0.1
1795	6.3	-6.7
1796	8.6	-6.6
1797	3.9	2.9
1798	2.4	3.9
1799	9.1	-4.4
1800	15.5	-12.9
1801	18.0	-11.7
1802	10.0	-6.8
1803	8.8	-4.2
1804	7.8	-0.6
1805	10.2	-4.9
1806	11.2	-5.0
1807	10.1	-3.4
1808	11.8	-8.6

Source: N. J. Silberling, 1919, "The British Financial Experience, 1790-1830," *The Review of Economic Statistics* (Amsterdam: Elsevier Science Publishers), Volume 1, No. 4, p. 287.

Tory ministry began to argue that the circumstances of the war itself, "[t]he avowal of our enemy to ruin our public credit,"[48] necessitated the continuation of restriction until after the commencement of peace. Once the Pitt ministry began to refer to the war with revolutionary France as a war of finance, and to restriction as a plan, to question it seemed less than patriotic. This was a very persuasive mode of argumentation.

During the period from 1797 to 1799, immediately following the suspension of payments, pamphlet and review comment was limited to the registration of approval or disapproval of suspension in principle. There was some comment on the role and responsiblities of the Bank of England, but the debate did not evolve into an analysis of the banking system as such. Radicals, such as Thomas Paine (1737–1809), in *The Decline and Fall of the English System of Finance* (1796), and many Whig reformers prophesied that inconvertible Bank of England notes must go the way that assignats had in France, bringing ruinous inflation.[49] Such dire consequences did not, however, materialize during this early phase. At this time there was a small premium on gold, but there was little depreciation of exchanges.[50] (See Table 3.5.)

Representative of the attacks on Pitt and the Tory government's policy of continued restriction was Edward Long Fox's *Cursory Reflections on the Causes, and some of the Consequences of the Stoppage at the Bank of England* (Bristol 1797). The only really concrete suggestion that went beyond the jeremiad against Tory-led restriction was a proposal to make Bank of England notes legal tender, so as to avoid any loss of confidence in their value. This legal tender question was recurrent in the controversy. The reluctance on the part of Pitt and the Tories to make Bank of England notes legal tender appears to have stemmed from no more than sensitivity to the jibes of Whig parliamentarians regarding the Tory's apparent imitation of the French monetary policy, which had already so conspicuously failed.[51]

In a measured response to Fox, Sir Francis Baring, the *de facto* leader of the London merchant/banking community, published *Observations on the Establishment of the Bank of England* (London 1797). In this pamphlet Baring really began the analysis of the banking system, arguing the question of the roles and responsibilities of the central and country banks as elements of the system at large.[52] Baring felt that country banks should not have the right of issue, and called on the government to make the Bank of England's notes legal tender. He warned that the popular distinction between government and private bank paper was unimportant, since any and all paper would depreciate when overissued.

In late 1799 there began a sharp rise in the price of foodstuffs, which continued into the nineteenth-century. (See Table 3.2 above.) The obviousness of the explanations of poor harvests, and of import shipping disruption due to war, forestalled comment linking price changes to the monetary situation at that time. Amongst the few exceptions to this was Thomas Malthus (1766–1834) who, in *An Investigation into the Cause of the Present High Price of Provisions* (London 1800), discussed just such a linkage. Importantly, however, Malthus saw the increase in money supply as the effect of higher prices, rather than as a/the cause.[53]

Beginning in late 1799 and continuing throughout 1800 there was also a marked depreciation of the pound on the Hamburg exchange, and the market price of gold rose well above its mint price. (See Table 3.5 above.) In early 1801 Walter Boyd published *A Letter to the Right Honorable Mr. Pitt, On the Influence of the Stoppage of Issues in Specie at the Bank of England, and on the Prices of Provisions and other Commodities* (January 1801). Boyd argued that restriction allowed the Bank of England to issue more paper than it would have been able to do under a system of convertibility. Having established no more than the possibility, Boyd, assuming the rule of profit motive, deduced the fact. He argued that overissue was the cause of exchange depreciation, of the rise in the market over the mint price of gold, and of rising commodity prices.

Boyd adhered to a deductive mode of argumentation; he assumed that the Bank of England directors would attempt to maximize profit. He exonerated country banks from any blame for overissue on the grounds that they still had to redeem their notes in Bank of England paper.[54] Boyd squarely blamed the Bank of England for all the troubles being experienced, and the Tories for allowing them to do it. He argued (deduced) that an extended period of exchange depreciation must be caused by overissue, since it would otherwise be adjusted for by increased export and decreased import receipts—trade and payments adjustments. This mode of deducing overissue from the possibility, and proving it from long-run exchange depreciation was a technique later to be associated with Ricardo's name. It was a characteristically bullionist and characteristically individualist argument in form and structure.

Boyd may be said to have had an axe to grind. He was a well-known figure at that time as both a member of parliament and a private banker. He had been the source of a sound argument against the restrictive credit policy of the Bank of England during 1796–97.[55] The business in which he was a partner, Boyd, Benfield and Company Bankers, failed in 1800 following the Bank of England's refusal to extend credit, and Pitt's refusal to allow them to contract for the government loan of 1799.[56] Pitt's refusal would appear to have been made in the light of an experience that was not then public knowledge.[57] Always a controversial figure, Boyd had every reason to be less than generous towards the Bank of England and the Tory government. His pamphlet, highly embarassing to both Bank and government, incited a wide response, some of which was dismissed by Boyd as the work of "subsidized writers."[58]

In a response to Boyd Sir Francis Baring's *Observations on the Publication of Walter Boyd* (London 1801) sliced past the polemic and hit at the weak point of Boyd's argument. He pointed out that the exchange changes were proportionately smaller than were the commodity price changes, and so could not be *the* cause of them.[59] Baring suggested that the actual situation was far more complex than Boyd's analysis allowed. Baring sought to treat each case on its merits and seemed to revel in the empirical complexity of things. He went into a detailed analysis of the facts of the case and took exception to Boyd's attempt to apply general principles. Baring's was a characteristically hierarchist mode of argument.

These are not merely substantively opposing arguments; they are opposing modes or styles of thought. They are, moreover, the styles of thought characteristic of the individualist and hierarchist positions in socio-cultural space. The Boyd-Baring exchange raised the crucial issue of the role and responsibilities of the Bank of England and of the country banks. Their analyses were alike insofar as they both placed the Bank of England in a pivotal position, but they disagreed on the *facts* and on how to arrive at them. Baring believed that in fact the Bank of England had not overissued, while Boyd apparently deduced that it had from the fact that it could, and would profit by doing so.

Sir Francis Baring was a leading member of the old order of bankers. Boyd, on the other hand, had recently experienced a sudden decline in power, in self-determination. Formerly a business leader, Boyd had been ruined by the actions and *rulings* of the Bank-Tory group. Boyd's experience of the factions must have increased as his ruin was brought on by Tory ministerial and Bank of England coordination subsequent to his anti-Bank critiques. Boyd's disgrace can be expected to have greatly increased his sense of exclusion, and thus his sense of group. His individualistic business life had brought him into conflict with the ruling elite(s) in parliament and the City of London and resulted in an exclusion that would make him conscious of the existence of groups or factions of power.

In terms of three-dimensional cultural schema Walter Boyd was, as a member of the emergent economic elite, positioned negative group—an individualistic entrepreneur. The circumstances of his ruin not only affected a decline in his regulative power, but also raised his experience of the groups. He was made painfully aware that there was a them and us. Hence in relation to his peers Boyd must be placed higher along the group dimension. Boyd was moving up group from the individualist negative region toward the zero position while slipping to a lower power, less self-determining position. (See Figure 3.4.)

The next stage of monetary debate arose in April 1802 with the renewal of the Restriction Act following the Peace of Amiens. Given that restriction was being portrayed by the Tories as a measure necessitated by war, the securement of peace should have brought an end to restriction, but it did not. Henry Addington[60] had replaced Pitt as prime minister and chancellor of the Exchequer in 1801, but there had been no real change of policy or of key power players. The ministerial changes were merely cosmetic; the government remained basically Pittite. Indeed Pitt returned as prime minister in May 1804.[61] The character of Henry Addington has been captured in the quote: "He was a unswerving old Tory. . . ."[62]

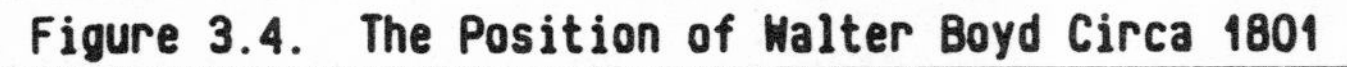
Figure 3.4. The Position of Walter Boyd Circa 1801

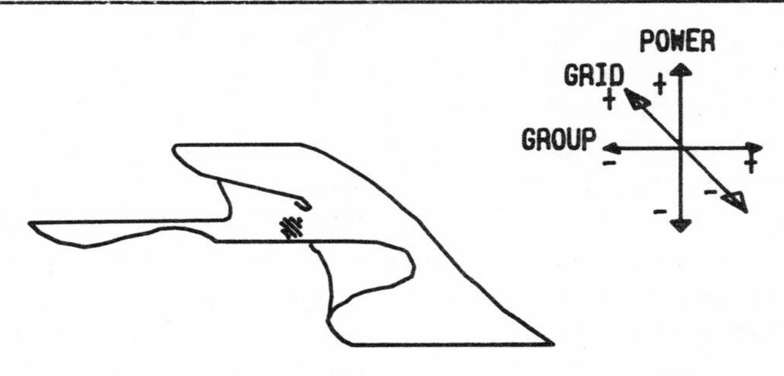

⋇ Indicates Position.

Source: M. Thompson, 1982, 'A Three-Dimensional Model,' In Mary Douglas, Ed., *Essays in The Sociology of Perception* (London: Routledge & Kegan Paul), p.50.

On the occasion of debating the renewal of the Restriction Act in the House of Commons on 9th April 1802, Addington expressed his conviction that the Bank of England could resume payments, and indeed was willing to do so.[63] Following Pitt's example Addington declared:

> The measure cannot furnish a pretence to the most timid man in the House, to suppose the Bank does not possess within itself the most ample means of satisfying the full extent of the demands which may be made on it.[64]

He suggested that restriction be continued until such time as exchanges returned to par or became favorable.[65] Addington also mentioned the pressure for commercial credit in the post-war climate, and the need to wait until they were able to judge the effects of peace before enacting resumption.[66]

Speaking in the House of Commons Mr. Jones mentioned a number of things about which, in the context of Addington's proposal to continue restriction, he was concerned. He was worried about the upsurge in forgery of Bank of England notes and about the number of executions for the offense; he was also worried about the "swarm of country banks," which, he said, had recently "spread like locusts."[67] This is exactly the kind of rhetoric of pollution and purity, of plague and health that is characteristic of high group commitment. Mr. Jones was a director of the Bank of England.

In the continuing debate on 21st April 1802 George Tierney[68] requested that steps be taken to ensure that the Bank of England made some real preparation for resumption. He remarked that it was odd that "peace and war were equally an argument for continuing restriction."[69] On the basis of these doubts Tierney successfully introduced a clause into the Continuation Bill, which allowed the Bank of England discretionary power to resume as and when it could. On that basis a one-year continuation of discretionary restriction into peacetime was carried.

It is notable that the rationale for the restriction of payments had undergone considerable change. Begun as a response to the financial panic and specie drains caused by the fear of invasion in 1797, restriction now apparently became a wartime strategy. After peace had come restriction continued. It was now being argued that to resume payments while exchanges were unfavorable would entail an external drain, and a return to the reserves crisis—a crisis that was not publicly admitted.

Henry Thornton[70] published his seminal *Enquiry into the Nature and Effects of the Paper Credit of Great Britain* (London) in 1802. In this work Thornton set out the responsibilities of a central bank, and began a complex and subtle analysis of the mechanisms of the banking system. He greatly raised the level of theoretical debate, but his work attracted little immediate attention. During his analysis Thornton separated the three main elements at issue: the market versus the mint price of gold, the exchange variations, and the price of provisions. However, it is perhaps the discursive impact of Thornton's *Paper Credit* that is of most interest.

Thornton took issue with Adam Smith on two counts. First, in regard to Smith's famous argument that the channel of circulation can never be overfull, since excess money would flow overseas and lead to the export of gold and to a rise in domestic prices, Thornton asserted that the channel of circulation can never be said to be full. Rather he argued that price inflation would absorb the extra money in the first instance.[71] The effect, according to Thornton, was that coin would be carried abroad in search of a better market. He had inserted an intervening variable, price rises, into Smith's schema. Second, he said that Smith had asserted that a permanent excess of the market over the mint price of gold was "entirely referrable to something in the state of the coin."[72] Thornton objected to the qualitative implication of this and sought to show that, barring "temporary extraordinary events,"[73] such a divergence was necessarily the effect of a quantitative excess. This had the effect of shutting off the equivocality involved in the qualitative versus quantitative elements of the interpretation of depreciation, but it also demanded a means of accommodating or reassessing the *facts*. In this it may have acted as a catalyst for the development of an important element of monetary theory, namely, the notion of the velocity of circulation.

Thornton was at pains to stress the importance of the velocity of circulation.[74] He asserted that no theory based on a simple quantity analysis would do. Thornton thought that prices were regulated by the demand *and* supply of the commodity, *and* the demand *and* supply of the means of payment: money.[75] In terms of the concrete situation of the time Thornton believed that fear of invasion had raised the demand for money—especially in the form of gold, because of the desire to hoard—and thereby the price of provisions. Therefore, Thornton saw the increase in money issue as the effect, rather than the cause of high prices. He did not, however, deny that under other circumstances it could be the cause. Interpretive flexibility was thus maintained during the development of the means to accommodate the facts with a quantity theory of money.

Inspired by Boyd's work,[76] Thornton argued, in contradistinction to Boyd, that the necessity for restriction and for its continuance had arisen from subsidy payments and from payments made necessary by poor harvests,[77] and that these payments abroad, rather than overissue from the Bank of England, were the cause of adverse exchanges in 1800–01.[78] As Francis Horner remarked at the time, it was the "one great object of [Thornton's] . . . book to persuade the public that there has been no such increase"[79] in note issue. Thornton made much of demand factors and of the velocity of circulation, showing that central bank control was in fact very limited. He downplayed the role of monetary factors in exchange and commodity price changes.[80]

Thornton also addressed the issue of the depressing effects of the restrictive Bank of England credit policy, the situation that Boyd had complained of. Indeed it has been said that "[t]he emphasis of his book of 1802 was on the consequences of ill-timed contraction."[81] Importantly, however, Thornton's analysis in this regard also worked in both directions. That is to say that it was equally applicable in the case of overissue, and in this guise it came to provide the bullionists with an argument against restriction. While Thornton was commonly perceived as a "champion of the Restriction bill,"[82] the fact that his analysis could be used to demonstrate a reverse relation tended to co-opt Thornton to the bullionist cause over time. Thornton's analysis of the mechanisms of the banking and monetary system was open to interpretive flexibility, but its mode—empirical complexity wherein the specific details of each case were examined in minutiae—was characteristically anti-bullionist in 1802. It was characteristic of hierarchist position of the old Tory/merchant elite. It followed the stylistic pattern introduced by Sir Francis Baring.

Indeed Thornton was a merchant and bank director, and his brother Samuel was a director, and sometime governor, of the Bank of England from 1780 to 1833. It is quite noticeable that in Thornton's *Paper Credit* there is something of a bias towards vindicating the activities of the Bank of England directors.[83] In concluding almost every chapter, especially towards the end, Thornton took occasion to remark on the value of the Bank of England in its present form, and to make a policy plea for hands off.[84] As Francis Horner remarked in a review of the work, "Mr. Thornton . . . employs several elaborate pages, to relieve the Bank from every degree of blame."[85] The *Monthly Reviewers* also noted that "the tendency of his arguments is in favor of the banking interest."[86] The thrust of Thornton's analysis was to show that prior to 1802 currency had *not* depreciated through overissue.[87]

Thornton did, however, implicitly criticize the Bank of England's commercial practices. He argued that the Bank should regulate its commercial loans with reference to the extent of circulation, as well as the wisdom of the individual loans. He wrote:

> [T]he Bank ought to regulate the total amount of its loans, with a view to the quantity of circulating medium, independent altogether of the solvency and opulence of those who wish to become borrowers, and the character of the Bills that are offered for discount.[88]

Despite Thornton's defense of the Bank directors he did, thereby, raise the basic issue of the role and responsibilities of the Bank of England. He was suggesting a new self-perception for the Bank as an organ of public policy, rather than as simply a private business. Perhaps it would be more accurate to say that he was pointing out that circumstances had pushed the Bank of England into a new role.

So what can be made of Henry Thornton's position? Born into a family of merchants engaged in Russian and Baltic trade, Thornton had been brought up in a relatively affluent, but very religious environment. He was a close friend of Wilberforce and a leading member of the Clapham sect. He was a partner in the bank of Downe, Free and Thornton from 1784 and member of parliament for Southwark from 1783 until his death in 1815.[89] Thornton was a spiritual, rather than a political reformer. The members of the Clapham sect were known by many in parliament as the "party of saints." He was in life as he was in parliament, a true independent. This would place him somewhere near the central position of the hermit.

Figure 3.5. The Position of Henry Thornton Circa 1802

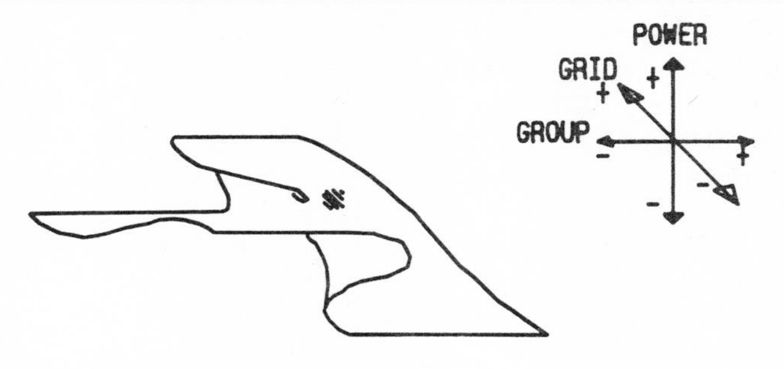

※ Indicates Position.

Source: M. Thompson, 1982, 'A Three-Dimensional Model,' In Mary Douglas, Ed., *Essays in The Sociology of Perception* (London: Routledge & Kegan Paul), p.50.

Thornton's work was somewhat equivocal as regards analytical content, but stylistically it was clearly anti-bullionist. His analysis was painstakingly detailed, and the underlying relationships that were posited were seen to operate differently depending on the particular circumstances of the case in hand. We can readily see how this style of thought, mode of analysis or cultural bias matches Thornton's social relations and life experiences. His involvement in a group such as the Clapham sect is clearly indicative of a high group position, his strict religious upbringing is indicative of a regulation in his life that would place him in a high grid position, and the long-term success of the family's mercantile business in the international arena is suggestive of a relatively high power position. Stylistically in the same region as the Tory/merchant elite and in large part enjoying that way of life, Thornton's deeply felt religious convictions and involvement in a group widely referred to as saints indicates an aloofness or other-worldly orientation that suggests he should be placed much closer to the central position of the hermit than his tory/merchant social peers—a relatively detached hierarchist position. (See Figure 3.5.)

Thornton's *Paper Credit* was quite favorably reviewed by Francis Horner[90] in the first issue of the newly founded *Edinburgh Review*. Given the enormous success of the *Edinburgh Review* Horner's review carried Thornton's work to a much wider audience than it would otherwise have had.[91] The Reverend Sydney Smith orchestrated the founding of the Edinburgh-based literary review in 1802. The co-founders were Francis Jeffery, Henry Brougham, and Francis Horner. All had been students of Dugald Stewart in Edinburgh, which, while no longer a political capital was still a major European social and intellectual center. The *Edinburgh Review* was an enormous success from the outset, being for some years the major publication of its kind and a significant platform for political economy. From its inception it appeared in the blue and buff colors of the Whig party, but it was not, at first, simply an organ of that party. As the founders drifted away toward political life in London, however, the review experienced a marked shift toward the rationalistic mode of discourse that became characteristic of reformist Whigs—circa 1807–09.[92]

In this first issue of the *Edinburgh Review* Horner congratulated Thornton on his impressive presentation of "well authenticated facts,"[93] but he criticized the order and manner of presentation of his arguments. In his attempt to present Thornton's argument in a better form Horner carried the analysis a little further.[94] Horner tried to clarify Thornton's analysis of the causal sequence in the relationship between currency and commodity price variations. Horner favored seeing higher commodity prices as the cause of higher gold and foreign bills prices when there was no movement of gold.[95] That is to say that he agreed with the direction of causation suggested by Malthus, and sought to limit the alternative interpretation that Thornton's analysis left open.

Horner also highlighted an important point of contrast between the analyses of Boyd and Thornton which was implicit in Thornton's choice of title, namely, that Thornton founded his analysis on a much broader definition of money than did Boyd. Indeed Thornton went so far as to say that "even in theory, no definite boundary can be marked between the circulating medium and the commodities of which it facilitates the exchange."[96] This is a crucial point in understanding the theoretical debate between the analysts of this perception, such as Thornton, Horner, and Malthus, and the more characteristic bullionists, such as Boyd, and Ricardo, who insisted on a more restricted definition of currency. It is a point to which attention is drawn by the use of the cultural theory approach suggested herein, because it reflects the styles of thought characteristic of position, namely, empirical complexity versus abstract simplicity. We shall return to this point in the examination of Ricardo's work.

After the year for which the Restriction Bill of 1802 had been enacted the scenario was unchanged. England was still at peace. Addington expressed his confidence in the solidity of the Bank of England and noted its readiness and willingness to resume specie payments, but since the exchanges were still against England such was not advised. Speaking to the Commons on 7th February 1803 Addington stated that "he did not think it necessary to enter into any inquiry respecting the sufficiency of the Bank to answer all the demands on it,"[97] but that unfavorable exchanges were sufficient reason to continue restriction. Fox wondered why the ministers mentioned only exchanges when their pamphleteers made so much of the reserves drain of corn imports and of loan and subsidy payments. He speculated that restriction and overissue may in fact be the *cause* of the unfavorable exchanges, and not a consequence.[98] Henry Thornton, for his part, was willing to agree with the chancellor that restriction must continue while the exchanges were unfavorable.[99]

During debate in the House of Lords on 22nd February 1803 Lord King[100] returned to the point that unfavorable exchanges were, according to his quantity theory of money, a *consequence* of excess issue.[101] This key relation was beginning to emerge in policy debate, but thus far it lacked theoretical foundation. Nevertheless, the period of peace had provided a turning point in the debate and in the thinking of a number of the analysts. The question of exchanges became linked to the continuation of restriction. Opponents of restriction began to argue that restriction was a/the cause of unfavorable exchanges. In this way the development of the theoretical element of causal direction can, like that of velocity of circulation before it, be traced back to the debate and its context. Perception of the direction of causal relations between restriction and overissue and exchange depreciation was a feature of perspective or cultural bias, *not* of the evidence.

By the time of the next parliamentary session, of course, an excuse for the continuation of restriction had presented itself as war with Imperial France had broken out. Parliament duly renewed the Restriction Act until six months after a definite treaty of peace had been concluded. On the occasion of the Commons debate on 30th November 1803 the chancellor of the Exchequer expressed the opinion that since the war had now resumed there would be no objection to the renewal of the Restriction Bill.[102] He once again testified to the strength of the Bank of England, saying, "It is satisfactory to know that the credit of the Bank of England had remained firm and unshaken."[103]

Irish Currency and Its English Fallout

The renewal of the English and Irish Restriction Acts in the (northern hemisphere) spring of 1803 provided the occasion for renewed monetary debate both inside and outside parliament. A pamphlet by Lord Peter King, a descendant of John Locke,[104] which derived from his speech in parliament in May, touched off this new phase of the debate. In *Thoughts on the Restriction of Payments in Specie at the Banks of*

Table 3.6. Bank of Ireland Notes at Dublin and Irish Versus English Exchanges

Year	Quantity Bank of Ireland Notes on Issue	(Pounds)	Exchange Percentage at Dublin
1797	Jan 1	621,917	2-1/3 per cent
	Apr 1	737,268	
	Jun 1	808,612	
	Sep 1	959,999	
1801	Apr 1	2,226,471	- 3-2/3 per cent
	May 1	2,405,214	
	Jun 1	2,350,012	
1802	Jun 1	2,678,980	- 3-2/3 per cent
	Aug 1	2,628,958	
	Oct 1	2,528,951	
	Dec 1	2,530,867	
1803	Feb 1	2,530,867	- 3-2/3 per cent
1804	Jan 1	2,986,999	- 8 per cent

Sources: J. R. McCulloch, 1966, Ed., *A Select Collection of Scarce and Valuable Tracts and Other Publications on Paper Currency and Banking* (New York: A. M. Kelley), p. 343; and F. W. Fetter, 1957, Ed., *The Economic Writings of Francis Horner in the Edinburgh Review 1802-1806* (London: University of London Press), p. 88.

England and Ireland (London 1803) King argued that inconvertible paper was deprived of its natural standard. He wrote that when

> a paper circulation cannot be converted into specie, it is deprived of this natural standard, and is incapable of admitting of any other. The persons to whom the duty of regulating such a circulation is entrusted are in danger, with the very best intentions, of committing perpetual mistakes.[105]

He went on to accuse the directors of both the banks of England and Ireland of overissue. King suggested that a fall in foreign exchange rates was proof of overissue if the change was not merely temporary, and/or the extent of the variation was greater than the real cost of transporting bullion. According to King a real exchange variation could never exceed "what will be sufficient to pay the expenses and the profit of the merchant who exports precious metals to restore the balance."[106]

In the context of the debate on Irish restriction King produced evidence which he believed proved depreciation by excess issue. He noted that by 1803 the number of Bank of Ireland notes in circulation had increased fourfold since 1797. (See Table 3.6.) King was by no means the only one to notice this increase. In the very influential weekly *Political Register* William Cobbett (1763–1835) wrote: "Bank-notes [are] in that happy country, a most flourishing branch of manufacture."[107] On the basis of his data and analysis King accused the directors of the Bank of Ireland of gross misconduct:

> An important trust, which, on mistaken principles of political necessity, was committed to this corporate body by Parliament, for the public benefit, appears to have been perverted to the private interest of the proprietors of their stock.[108]

He went on to note that "[a] power has been committed to the directors of the Bank, which is not entrusted by the constitution even to the Executive Government."[109] Addressing himself specifically to Irish currency King recommended that the Bank of Ireland be made to redeem its notes in Bank of England notes.[110] This, rather oddly, would have centralized power with the Bank of England, rather than with government.

When Francis Horner reviewed King's pamphlet in the *Edinburgh Review* he stated his wish to précis the work in order to diffuse correct opinion on the laws of currency[111]—that is, to push the popular bullionist line, rather than to split analytical hairs. He concluded with little but praise for Lord King, and remarked that this effort might be a model for pamphlets in the future—a comment which Ricardo may have noted, though the pamphlet was *not* amongst Ricardo's private library.[112]

The argument that Horner presented in the review was couched in terms of the unfairness of either inflation or deflation. He concentrated on the commodity price effects of monetary policy, and he sought the maintenance of stable price levels in the interests of distributive justice. The distinction between this and Ricardo's bullionism will become clear. Horner also noted that Lord King followed the basic argument of Boyd's earlier work, "setting out with the strong presumptions which might have led us to expect an excessive issue of paper, and confirming that probability by reference to the price of bullion and the rate of foreign exchanges."[113] That is to say that King adopted the style and mode of analysis of Boyd, a style characteristic of bullionists/characteristic of individualists. Like Boyd, King sought to reduce the situation to one of a few simple relationships, and to uncover a causal mechanism.

What can be made of Lord King, apparently an aristocratic landowner, exhibiting the stylistic and substantive features of bullionism—the cultural bias of the new order? Lord Peter King had a long family history of Whig liberalism; he was thoroughly immersed in an opposition to high Toryism. Having a background in and the economic means to power, Lord King can be placed high on the power dimension. It is clear from his family background, however, that he did *not* share

Figure 3.6. The Position of Lord King Circa 1803

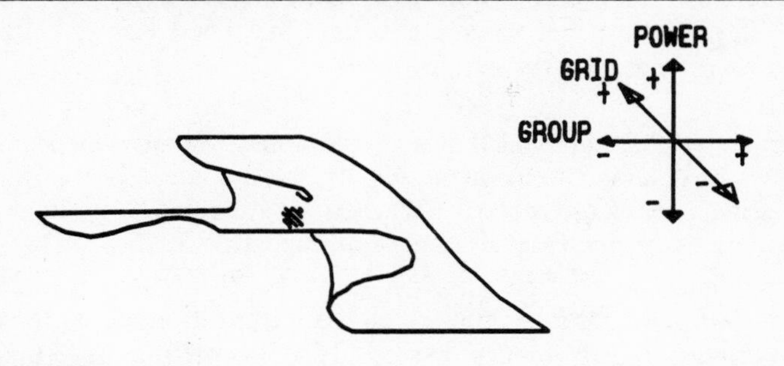

* Indicates Position.

Source: M. Thompson, 1982, 'A Three-Dimensional Model,' In Mary Douglas, Ed., *Essays in The Sociology of Perception* (London: Routledge & Kegan Paul), p.50.

the Tory sense of order; he was not hamstrung by a web of conventions and obligations. He can, therefore, be placed low negative grid. Power gave him a relative autonomy. Though educated in the "right places," and in the aristocracy, King's Whig liberalism placed him on the outer edge. He is, indeed, reported to have been something of a recluse.[114] This clearly suggests an experience of individual exclusion; and a placement in the negative group region. In addition to this broad siting, however, Lord King was in a rather unique position. He was educated amongst, and moved amongst the old Tory elite. His daily life experience, his relatively privileged position, brought him into regular contacts which must have raised his awareness of the existence of groups, of the factional them and us. He is, in short, to be placed in a high negative group position. It is this which Lord King had in common with Walter Boyd, and in this that one can identify as the basis of their shared style of thought—their bullionism. (See Figure 3.6.) Though different in terms of specifics their life experiences put them in a similar position regarding their pattern of social relations, hence they developed and deployed a similar cultural bias, a cultural bias characteristic of individualists.

It is important to note here that a conventional class-based or simple pecuniary or social interest-based analysis would flounder on this, where cultural theory succeeds. Lord King is not the anomaly he would appear to be for conventional historiographic methods, but rather the exception that proves the rule for cultural theory. Employing a cultural approach reveals the clear and predicted correlation of social relations and style of thought—of social experience and cultural bias.

A key feature of Horner's critique concerned King's reliance on facts. Horner referred to the situation as

> a case of that sort which scarcely admits of direct evidence. Even when we can obtain faithful documents of the amount of notes in circulation during a series of successive periods, we are not fully entitled to consider a progressive increase of amount as conclusive.[115]

This was so because the rates of circulation and the quantity of commercial requirements varied. Nor, according to Horner, did high and rising prices necessarily indicate overissue. Horner went on to say that there were two "very simple and satisfactory tests, by which the fact of an excessive currency may be ascertained,"[116] namely, an excess of the market over the mint price of gold beyond a temporary imbalance, and a fall in foreign exchanges which is longstanding and/or exceeds the real cost of transporting precious metals. Nevertheless, Horner qualified this to some extent by enumerating other possible causes of these two circumstances. He left it unclear as to the degree of certainty achieved by these indicators, effectively suggesting that they were necessary but not sufficient conditions.

Horner did nothing to reduce interpretive flexibility or equivocality, because he used both the velocity of circulation argument and raised doubts over the meaning of gold price and exchange variations. Francis Horner broadly agreed with the bullionist argument, but he did not do so simplistically, nor did he do so stylistically. His caveats suggest an appreciation of empirical complexity, and an uncertainty as regards key causal relations.

During debate in the House of Lords on 3rd May 1803 Lord King spoke of the gross overissue indulged in by the directors of the Bank of Ireland. At the same time he excused country banks from all blame for the excess issue.[117] Subsequent parliamentary debate on Irish Bank restriction spanned three readings of the Bill from 9th December 1803 to 20th February 1804. During discussion Lord King noted that by restriction the interests of the Bank had been placed in opposition to its national duty. There was, nevertheless, general agreement that it was not possible to lift restriction at the Bank of Ireland without also doing so for the Bank of England. Lord Archibald Hamilton stressed the centrality of the Irish/English exchange parity,[118] and suggested that a limit for Bank of Ireland note issue be legislated. In response Lord Henry Petty pointed out that the quantity of issue was not a sufficient criterion to judge excess. He drew attention to the extent to which the Bank of Ireland had gained power through restriction, noting that the Bank had increased its dividends some 5 to 6½ per cent.[119]

While the relative English/Irish exchanges and the role and responsibility of a central bank were uppermost in debate in the Lords, in the Commons Thornton shifted the focus onto the circumstances of other private banks in Dublin. Such banks shared with the Bank of Ireland the right to issue in the metropolis. In view of this situation he suggested that the Bank of Ireland did not have the same control over the money supply that the Bank of England did in London, and could not carry all the blame. Nevertheless, Thornton concluded that the Irish exchange problem was clearly caused by "excess paper circulation."[120]

In March 1804 the House of Commons established a committee of enquiry to investigate the state of the Irish currency. Henry Thornton played a major role.[121] Hearings of the Irish Currency Committee began early in March 1804 and are said to have consisted of questions arranged largely by Henry Thornton.[122] The *Irish Currency Report*[123] was submitted to the House of Commons on 19th June 1804, and it concluded "that this depreciation in Ireland arises almost entirely, if not solely, from an excess of paper."[124] The committee recommended that the Bank of Ireland take steps to redeem its paper in Bank of England notes or London funds. Interestingly the report was not discussed in parliament, though an attempt to do so was made in April 1809.[125]

Those with Bank of Ireland and/or administration interests or connections argued that it was not, or not only, overissue of bank paper that had caused depreciation. They suggested that remittances to England—in the form of Irish loans payable in England, or of rents payable to absentee landlords also played an important part. Those connected with landlord interests in Ireland, such as Sir Henry Parnell, argued, conversely, that remittances were a minor problem compared to Bank overissue. While Dublin bankers were "accumulating immense fortunes by the unrestrained and arbitrary issue of paper"[126] absentee landlords were being hurt because their rents were of decreasing Bank of England paper value. It was this scenario that lay behind the Irish currency question. It is a marked point of contrast with later debate, and an interesting sidelight on discursive evolution, that the issue was not so much the return of Irish paper to its former value in gold, but was rather the need to return Irish paper to its English *paper parity*.[127] In this phase of the debate it was all eyes on exchanges, on paper parity, with little mention of the price of gold.

Table 3.7. Annual Average Amount of Bank of England Notes in Circulation, 1792-1822

Year	Value Pounds (millions)	Year	Value Pounds (millions)
1792	11.4	1808	17.1
1793	11.6	1809	18.9
1794	10.8	1810	22.5
1795	11.5	1811	23.3
1796	10.2	1812	23.2
1797	11.0	1813	24.0
1798	12.6	1814	26.9
1799	13.5	1815	26.9
1800	15.1	1816	26.6
1801	15.8	1817	28.2
1802	16.7	1818	27.2
1803	16.5	1819	25.1
1804	17.4	1820	23.9
1805	16.9	1821	21.6
1806	16.8	1822	17.9
1807	16.7		

Source: N. J. Silberling, 1919, "The British Financial Experience, 1790-1830," *The Review of Economic Statistics* (Amsterdam: Elsevier Science Publishers), Volume 1, No. 4, p. 291.

Ireland and England

While it was unrealistic to blame only the Bank of Ireland, King's analysis of the Irish situation in general was persuasive, and was supported by a common-sense interpretation of the evidence at hand. However, the simple extension of this argument to the situation in England was far from convincing. The overissue argument in regard to the Irish currency problem was considerably stronger than that in regard to English currency, although few analysts drew a clear distinction between the two cases at the time. King's simple quantity theory was adequate to the task in the Irish case, but was rather weak in the English case. Compare the rapid issue increase in Ireland 1797–1804 in Table 3.6. (above) with the modest increase in England circa in 1800–1808. (See Table 3.7.) The English case obviously needed something more than simple quantity theory. So it is no surprise to find that further development of the velocity of circulation argument awaited debate on English currency.

In the context of the English bank restriction Francis Horner seemed quite willing to take over King's conclusions concerning Ireland lock-stock-and-barrel as a case against the Bank of England in 1803. Others followed suit. John Wheatley (1772–1830) weighed into the debate in this vein with some *Remarks on Currency and Commerce* (London 1803), and later added *An Essay on the Theory of Money and Principles of Commerce* (London 1807). Wheatley erected a quite extreme bullionist argument, stating that depreciation was due solely to monetary expansion, that this had only price and no production effects, and that foreign payments had no exchange effects.[128] In prescriptive terms Wheatley followed the line suggested by Adam Smith in the *Wealth of Nations* in relation to the Seven Years War, the argument being that wars can be financed by exports, with a convertible currency, provided that domestic prices are kept low enough for remittances to take the form of commodity exports.[129] Wheatley's argument exhibited a deductive style, the simple relations of quantity theory underpinned his analysis, and the particulars of the case, be it Irish or English depreciation, were unheeded. It was a characteristically bullionist argument in content and in style.

Wheatley came from a landed family of Erith, Kent, and was educated at Oxford. However, an explanation for his bullionist argument is apparent in his biographical details. He was associated, as something of a dependent, with the Grenville family, who were leading Whig liberals. He appears to have been in constant financial difficulties. The connection with the Grenvilles was close and may well have been stylistically influential. Lord Grenville (1759–1834) was later said by J. L. Mallet to "push the principles of political economy of the present school as far as Ricardo himself."[130] Wheatley lived a wanderer's life in South Africa and India. His stylistic bent might best be explained by his lifelong financial adversity, and his dependence on, and association with, Grenville.[131] Like Lord King, Wheatley was excluded, and had reason to be acutely aware of his position. He was a wanderer, but something of a dependent.

Such a set of social relations places Wheatley in the negative group region and in a relatively low power position—he was excluded from the mainstream and rather dependent. He was neither regulating nor, so long as he was indulged by the Grenvilles, regulated. In short, Wheatley can be placed in the same region in socio-cultural space as Lord King and Walter Boyd: just inside the individualist region, but close to the central hermit region. Like King he was a wanderer, a loner, something of a hermit. Little wonder then that Wheatley adopted the same mode of thought and style of analysis as had Lord King—that he had the same cultural bias. (See Figure 3.7.)

A rather unfavorable review of Wheatley's 1803 pamphlet, penned by Henry Brougham,[132] appeared in the *Edinburgh Review* of October 1803.[133] Brougham wrote: "Mr. Wheatley's errors and inaccuracies alone [warrant] the praise of originality."[134] The major point of contestation

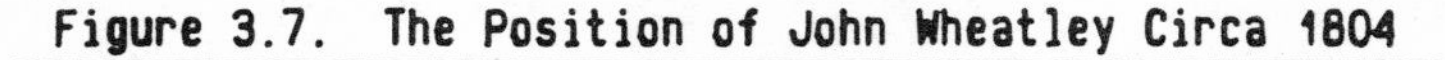

Figure 3.7. The Position of John Wheatley Circa 1804

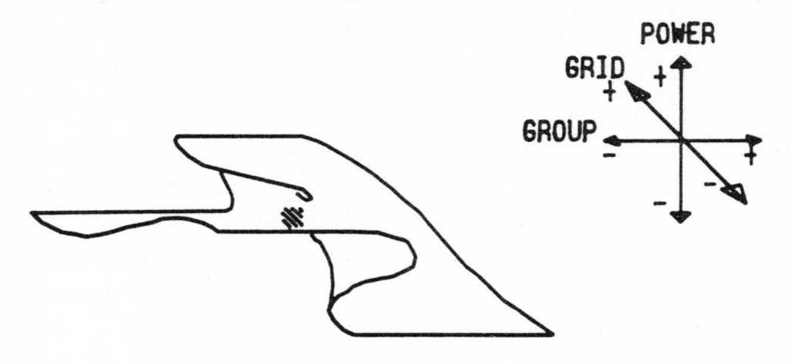

⁂ Indicates Position.

Source: M. Thompson, 1982, 'A Three-Dimensional Model,' In Mary Douglas, Ed., *Essays in The Sociology of Perception* (London: Routledge & Kegan Paul), p.50.

was Wheatley's use of the facts.[135] Wheatley based his whole analysis of the burden of public debt on a table of figures presented by Sir G. Shuckburgh to the Royal Society in 1798. With careful examination Brougham revealed that "some inferences may be drawn from this table, which are, in our apprehension, equivalent to a *reductio ad absurdum.*"[136] Since such was his foundation the reviewers felt inclined to dismiss all of Wheatley's conclusions, and to lament his lack of thoroughness and originality. Wheatley's analysis, such as it was, was adjudged guilty by association.

This review reveals the stylistic disparity between Wheatley and Brougham. The former relied on simple relationships and deployed facts supportively, while the latter was so impressed by the empirical complexity and specificity of the case that he failed to make note of the relations at all. Brougham debated only the facts, he did not take issue with the causal relations. Wheatley's style of thought, emergent bullionism, was strange to Brougham who was concerned exclusively with facts and displayed little appreciation of principles.

A singular and interesting point of contrast between Wheatley's analysis and that of some other bullionists was that he attributed monetary expansion to the unregulated country banks.[137] The key element, for Wheatley, was the issue of low denomination notes and coin. He virtually reversed the order of Boyd's analysis in his 1807 *Essay,* claiming that the Bank of England's issue of large denomination paper was in great part determined by the "quantity of inferior means of payment."[138] Wheatley recommended that the country banks be deprived of the right of issue, and that there be no issue of small denomination paper.

This suggestion was rather impractical, because the silver of coin for small change had, like gold, almost completely disappeared from circulation. Consequently there was a genuine need for small denomination paper. Nevertheless, Wheatley's views on country banks were shared by many writers who did not accept the bullionist doctrines. Malthus's 1803 edition of the *Essay on Population,* for example, suggested that local overissue (issue in local country areas) might raise prices throughout the country, once absorbed into circulation.[139] That is to say that Malthus, like Wheatley, pointed the finger at private banks rather than the central banks.

It was not, therefore, necessarily the case that the bullionists all exonerated the country banks and blamed the Bank of England. Wheatley differed from Boyd and King on this point. Given that it was not a necessary element of the bullionist argument, one must ask why the Bank of England was attacked by the majority of bullionists. The reason for Boyd's attack has already been hinted at, and the general perception of the Bank in the City community has also been noted. These reasons clearly relate to growing localized factionalism, and demonstrate the importance of context in the determination of the substantive detail of monetary analysis.

As has been noted, unlike Boyd and King, Wheatley was not directly involved in the financial community in the City of London; he was from the country. Hence the substantive difference of focus, and side, taken in the country bankers versus Bank of England struggle. The underlying social relations of power—relations of City factionalism and wartime political divisions—are clearly the foundation of the various perceptions of those engaging in the debate regarding the development of monetary policy and the development of monetary theory. There is a clear and predictably patterned relationship between social relations and cultural bias, between life experience and the mode, style and content of thought and its expression.

By the end of 1803 monetary debate had died down. Pamphleteers and reviewers alike seemed to have run out of things to argue about or of the will to argue, although it is clear that the debate had not been in any sense resolved. The *Irish Currency Report,* submitted to parliament in June 1804, fell dead from the press. There was suddenly no interest, no debate, no controversy.

AN ANALYTICAL SUMMARY

The Bullion Controversy of 1797 to 1808 followed the course that cultural analysis would suggest. For the duration of the war against revolutionary France (until 1802) those polarized over the war argued in a systematically and identifiably different manner about the financial means of its prosecution. As their respective positions polarized, identifiable and distinct styles of thought emerged, and there was an increasing drive to argue for their perspectives. Hence debate increased.

As events unfolded during the reign of terror in France, and Napoleon began an imperialist expansion (1803 to 1807–08), English men and women became more united. Their everyday social experience 1803–07 was unifying rather than divisive. Division gave way to patriotic unity; conflict gave way to consensus. The political manifestation of this was the coalition Ministry of All Talents 1806–07; the theoretical/discursive manifestation of it was the years without heated debate about currency 1804–07. The war with imperial France changed the mood and the minds of monetary theorists, by changing their social environment.

In the scheme of things the Irish currency controversy appears to have been something of a separate problem. It was a clash between landlords, merchants and administrators with Irish connections. It could perhaps best be seen as a situation of power groups establishing a balance of power after the Act of Union with Ireland in 1801. Basically the landlords and merchants affected an administrative change which ensured that the administration officials were not more favorably treated than they. The payment of Irish administration officials at English par was stopped.[140] Once this equity was restored, debate ceased, regardless of the extent of depreciation. The Irish currency debate was about Irish-English exchange balance, that is, paper equity, rather than about the value in gold of the Irish pound.

The relation between social environment and both bullionist and anti-bullionist styles of thought is clear. The analyses of Walter Boyd and Lord King led the way as regards the characteristic abstract, simplifying style of the bullionists/individualists. Sir Francis Baring gave witness to the contrasting empirical style of the old order/hierarchists, as did Henry Brougham in passing. The situations of Henry Thornton and Francis Horner were more equivocal. Their social position was somewhere between the two extremes, and their analyses reflected this. John Wheatley's analysis had many bullionist elements, but his peripherality as regards the City of London power plays placed him in a position from which a non-accusative stance toward the Bank of England was possible. This reveals the extent of the relation between

Figure 3.8. The Positions of Key Monetary Theorists (1797-1803)

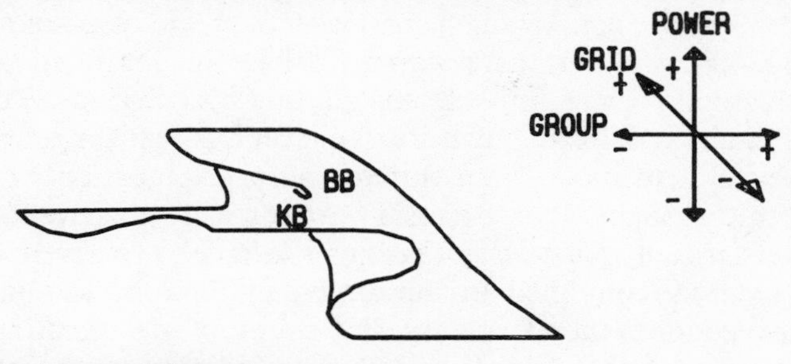

Where BB = Baring and Brougham
and KB = King and Boyd

Source: M. Thompson, 1982, 'A Three-Dimensional Model,' In Mary Douglas, Ed., *Essays in The Sociology of Perception* (London: Routledge & Kegan Paul), p.50.

social environment and both substantive and stylistic elements of monetary theory.

Mapping onto the three-dimensional cultural schema Lord King and Walter Boyd can be placed at the high end of the negative group position, approaching zero, while Baring and Brougham might be placed at the lower end of the positive group position. (See Figure 3.8.) This placement of the latter two is appropriate because both Baring and Brougham were rather peripheral to the old-guard Tory elite, and both were in a position to be aware of the increasing importance of the economic order. Just as one would expect, as the positions polarized (up to 1802) distinct styles of thought emerged from those in distinct positions, and the stragglers in the polarization—those who remained closest across, and to, the center—were the ones who were left talking to each other. These were the mediators, in the middle, representing extremes who could no longer speak the same language. This position lies at the heart of explaining how and why certain participants came to be regarded as central figures in the debate, while others did not. We shall pursue this at much greater length, but, in short, the spatially central are central.

A further important lesson to be drawn from the briefly debated Irish currency question is the performance of the cultural theory approach we have adopted in comparison to other traditional historiographic methods.

The timing of the *Irish Currency Report* and its stillbirth, as well as the death of pamphlet debate circa 1804, should not be seen as a simple result of the concrete fact of the depreciation of Irish currency—of events. In January 1816 the Irish pound reached a level some 18 per cent below par, as much or more than the maximum depreciation of 1803–04, and yet there was no debate or comment at that time.[141] The methodological implications of this are obvious. Positivist historians characteristically identify the event of exchange depreciation as the key. But to move from this to causal explanation in the history of economic thought, it is necessary to correlate the event (depreciation) with discussion. *This did not occur in 1816.* The two were not correlated, or constantly conjoined in the jargon of philosophers of science. Historians cannot, therefore, use the event of depreciation as a causal explanation of the debate, and thereby of the emergence of bullionist monetary theory. An alternative historiographic approach is necessary if we are to explain the course of the controversy regarding Irish currency, and in adopting a cultural theoretic approach we believe we are offering a superior alternative.

The task of the next chapter is to account for the upsurge in controversy circa 1808–10, and to analyze the character and content of the emerging alternative monetary theories during that period.

NOTES

1. A. Smith, 1776, *An Inquiry into the Nature and Causes of the Wealth of Nations,* 2 Volumes (London: Strathan and Cadell), and D. Ricardo, 1819, *On the Principles of Political Economy and Taxation,* 2nd Edition (London: John Murray).

2. T. W. Hutchison, 1978, *On Revolutions and Progress in Economic Knowledge* (Cambridge: Cambridge University Press), p. 294.

3. William Pitt (1759–1806). Younger son of the Earl of Chatham, he became a member of parliament in 1781, Chancellor of the Exchequer in 1782, and Prime Minister in 1783.

4. A. Briggs, 1959, *The Age of Improvement, 1784–1867* (London: Longman), p. 84.

5. G. D. H. Cole, 1971, *The Life of William Cobbett* (Westport, Conn: Greenwood), p. 120.

6. A. W. Palmer, 1964, *The Penguin Dictionary of Modern History, 1789–1945* (Harmondsworth: Penguin), p. 257.

7. D. Weatherall, 1976, *David Ricardo. A Biography* (The Hague: Martinus Nijhoff), p. 29.

8. J. E. Cookson, 1982, *The Friends of Peace; Anti-War Liberalism in England 1793–1815* (Cambridge: Cambridge University Press).

9. Briggs, 1959, op. cit., p. 124.

10. Ibid., p. 140.

11. Charles James Fox (1749–1806). Second son of Henry Fox, Lord Holland. He joined the Rockingham Whigs in 1774, and after the defeat of his India

Bill in 1783 he was relegated to a long role in opposition as the dualist of the younger Pitt.

12. Briggs, 1959, op. cit., p. 161.

13. Ibid., p. 162.

14. P. Mathias, 1969, *The First Industrial Nation* (London: Methuen), p. 44.

15. Such actions were sporadic throughout the period of the Combination Laws. The forms of reaction varied over time and place, and to some extent with the aim. Refer to John L. and Barbara Hammond, 1978, *The Village Labourer* (London: Longman); E. P. Thompson, 1963, *The Making of the English Working Class* (London: Gollancz); M. I. Thomis, 1972, *The Luddites; Machine Breaking in Regency England* (New York: Schocken); E. J. Hobsbaum and G. Rudé, 1973, *Captain Swing* (Harmondsworth: Penguin); etc.

16. Briggs, 1959, op. cit., p. 170.

17. N. J. Silberling, 1924, "The Financial and Monetary Policy of Great Britain during the Napoleonic Wars," 2 Parts, *Quarterly Journal of Economics,* Vol. 38, p. 219.

18. J-B. Say, 1964, *A Treatise of Political Economy, or the Production, Distribution and Consumption of Wealth* (New York: A.M. Kelley), p. LIV.

19. D. Winch, 1978, *Adam Smith's Politics; An essay in historiographic revision* (Cambridge: Cambridge University Press).

20. F. W. Fetter, 1965a, *The Development of British Monetary Orthodoxy 1797–1875* (Cambridge, Mass: Harvard University Press), p. 6.

21. Ibid., p. 34.

22. F. W. Fetter, 1942, "The Bullion Report Reexamined," *Quarterly Journal of Economics,* Vol. 56, p. 663.

23. E. Cannan, 1925, *The Paper Pound of 1797–1821* (London: P.S. King and Son) p. vii.

24. Francis Baring, 1802, quoted Fetter, 1965a, op. cit., p. 11. See also C. Cook and J. Stevenson, 1980, *British Historical Facts 1760–1830* (London: MacMillan Press), p. 188.

25. S. R. Cope, 1942, "The Goldsmids and the Development of the London Money Market during the Napoleonic Wars," *Economica,* Vol. 9, p. 202.

26. Gradually the hand of the Jobbers was strengthened by the rising wartime expenditure producing a falling market on trend.

27. E. V. Morgan and W. A. Thomas, 1962, *The Stock Exchange; its History and Functions* (London: Elek Books), p. 48.

28. E. T. Powell, 1916, *The Evolution of the London Money Market, 1385–1915* (London: Financial News), p. 158.

29. Weatherall, 1976, op. cit., p. 21.

30. Cope, 1942, op. cit., p. 181.

31. Morgan and Thomas, 1962, op. cit., p. 68.

32. Ibid., pp. 68–69.

33. Ibid., p. 70.

34. Fetter, 1965a, op. cit., p. 12.

35. Silberling, 1924, op. cit., p. 402; and F. W. Fetter, 1959, "The Politics of the Bullion Report," *Economica,* Vol. 26, p. 102. See also J. R. McCulloch,

Ed., 1966, *A Select Collection of Scarce and Valuable Tracts and other Publications on Paper Currency and Banking* (New York: A. M. Kelley), p. 446.

36. Cannan, 1925, op. cit., p. vii.
37. Silberling, 1924, op. cit., p. 398.
38. Ibid., p. 399.
39. Cannan, 1925, op. cit., p. xvii.
40. Fetter, 1965a, op. cit., p. 28.
41. *Quarterly Review,* 1811, Vol. V, Art. xi, p. 252.
42. Fetter, 1942, op. cit., pp. 663–65.
43. Silberling, 1924, op. cit., p. 418 and p. 423.
44. Fetter, 1965a, op. cit., p. 28.
45. Fetter, 1942, op. cit., p. 661.
46. McCulloch, 1966, Ed., op. cit., p. 96.
47. Cannan, 1925, op. cit., p. xviii.
48. William Pitt, quoted in *Hansard's,* Vol. XXXIII, p. 1030.
49. Silberling, 1924, op. cit., p. 401. Interestingly Paine had begun to link inflation and/or deflation with redistribution. He pointed out that some had the ability to pass on higher prices while fixed income (wage) earners could not. It was just such an awareness and argument, for a wage rise for his fellow Excisemen that ended Paine's Excise Board career. See H. Collins, 1969, Ed., *Thomas Paine; The Rights of Man* (London: Pelican), pp. 14–15. It was also one of the first expressions of the wage versus profit relation in the monetary context.
50. Fetter, 1965a, op. cit., p. 26.
51. Ibid., p. 30.
52. Ibid., p. 29.
53. T. R. Malthus, 1800, *An Investigation into the Cause of the Present High Price of Provisions* (London), p. 14., cited in Fetter, 1965a, op. cit., p. 30.
54. Silberling, 1924, op. cit., pp. 404–05, and Fetter, 1965a, op. cit., p. 31.
55. Fetter, 1965a, op. cit., p. 31.
56. Fetter, 1959, op. cit., p. 102.
57. The Boyd Benfield company was the subject of a later inquiry over the application by Lord Melville to the then Chancellor Pitt in 1799 for 40,000 pounds to bail out the company who were the holders of the government loan of 1796. In June 1805 parliament passed an Indemnity Bill for Pitt whose actions, though not legal, were adjudged necessary. Refer; *Hansard's,* 1805, Vol. V, p. 397f.
58. Fetter, 1965a, op. cit., p. 32.
59. Idem.
60. Henry Addington (1757–1844). Created Viscount Sidmouth in 1805. Entered the Commons in 1793 and was a close friend of Pitt. Speaker of the House 1798 to 1801, and Prime Minister 1801–1804.
61. *Political Register,* 1810, Vol. 18, No. 36, p. 1156.
62. Palmer, 1964, op. cit., p. 17.
63. *Political Register,* Vol. 11, No. 36, p. 1157.
64. *Hansard's,* Vol. XXXVI, p. 541.

65. *Political Register,* op. cit., p. 1158.
66. *Hansard's,* Vol. XXXVI, p. 543.
67. *Hansard's,* Vol. XXXVI, pp. 543–44.
68. George Tierney (1761–1830). Born in Gibraltar into an eminent merchant family of longstanding. Educated at Eaton and Cambridge. Tierney often clashed with Pitt; indeed they fought a dual before a large crowd on Putney Heath on 27th May, 1798. Tierney was Treasurer of the Navy under Addington 1802–04.
69. *Hansard's,* Vol. XXXVI, p. 548.
70. Henry Thornton (1762–1815). Banker, member of parliament for Southwark, and member of the Clapham sect of Humanists and Evangelicals. He was founder of the Sierra Leone Company.
71. Francis Horner, in F. W. Fetter, 1957, Ed., *The Economic Writings of Francis Horner in the Edinburgh Review, 1802–06* (London: London University Press), p. 35. See also B. Fontana, 1985, *Rethinking the Politics of Commercial Society: The 'Edinburgh Review' 1802–1832* (Cambridge: Cambridge University Press).
72. Henry Thornton, 1802, in McCulloch, 1966, Ed., op. cit., p. 270.
73. Ibid., p. 289.
74. Ibid., p. 139.
75. Ibid., pp. 262–63.
76. Horner, quoted Fetter, 1957, op. cit., p. 34. See also C. F. Peake, 1978, "Henry Thornton and the Development of Ricardo's Economic Thought," *History of Political Economy,* Vol. 10, No. 2, p. 194, and Fetter, 1959, op. cit., p. 103.
77. H. G. Grubel, 1961, "Ricardo and Thornton on the Transfer Mechanism," *Quarterly Journal of Economics,* Vol. 75, p. 293.
78. Peake, 1978, op. cit., p. 195. Refer also to T. Asromourgos, 1988, "The Life of William Petty in relation to his economics: a tercentenary interpretation," *History of Political Economy,* Vol. 20, No. 3, pp. 337–356, regarding the development of the idea of the velocity of circulation.
79. Horner, quoted Fetter, 1957, Ed., op. cit., p. 34.
80. D. A. Reisman, 1971, "Henry Thornton and Classical Monetary Economics," *Oxford Economic Papers,* Vol. 23, p. 86.
81. Fetter, 1965a, op. cit., p. 41.
82. *Quarterly Review,* 1809, Vol. 3, No. 5, Art. XII, p. 154.
83. McCulloch, 1966, Ed., op. cit., p. xii.
84. See, for example, Thornton, in McCulloch, 1966, Ed., op. cit., p. 209, p. 236, p. 278–9, p. 296, p. 325, etc.
85. Horner, 1802–03, *Edinburgh Review,* Vol. I, p. 197.
86. *Monthly Review,* 1802, Vol. 38, Art. IX, p. 30.
87. *Quarterly Review,* 1810, Vol. IV, No. VIII, Art. X, p. 425.
88. Thornton, quoted Horner, in Fetter, 1957, Ed., op. cit., p. 150.
89. L. Stephen and S. Lee, 1917, Eds., *The Dictionary of National Biography* (Oxford: Oxford University Press), 22 Vols., and F. A. Von Hayek, 1939, Ed., *An Enquiry into the Nature and Effects of the Paper Credit of Great Britain* (London: George Allen and Unwin).

90. Francis Horner (1778–1817). Scottish lawyer and member of parliament he was a co-founder of the *Edinburgh Review* in 1802. Horner entered parliament as the Whig member for St.Ives, as a nominee of Lord Landsdowne, and he quickly became the voice of political economy in parliament.

91. Fetter, 1959, op. cit., p. 103.

92. Refer; F. W. Fetter, 1953, "The Economic Articles in the Edinburgh Review and their Authors 1809–52," *Journal of Political Economy,* Vol. 61, pp. 232–59; J. Clive, 1957, *Scotch Reviewers: The Edinburgh Review 1802–1815* (London: Faber and Faber); P. Sraffa and M. H. Dobb, 1951–55, Eds., *The Works and Correspondence of David Ricardo* (Cambridge: Cambridge University Press), Vol. VII, p. 246; etc.

93. *Edinburgh Review,* 1802–03, Vol. 1, Art. XXV, pp. 172–201.

94. Silberling, 1924, op. cit., p. 411.

95. Ibid., p. 412.

96. Thornton, quoted Horner, in Fetter, 1957, Ed., op. cit., p. 31.

97. *Hansard's,* Vol. XXXVI, p. 1147.

98. Ibid., p. 1151.

99. Ibid., p. 1155.

100. Lord Peter King (1776–1833). Seventh Lord King, Baron of Ockham. Followed in a long family tradition of Whig liberalism, he supported religious toleration and opposed restriction and the corn laws.

101. *Hansard's,* Vol. XXXVI, p. 1157.

102. *Hansard's,* 1804, Vol. I, p. 1572.

103. Idem.

104. Cannan, 1925, op. cit., p. xxvi.

105. Lord King, 1803, quoted in *Monthly Review,* 1803, Vol. 41, Art. XI, p. 313.

106. Lord King, 1803, quoted Horner, in Fetter, 1957, Ed., op. cit., p. 85.

107. W. Cobbett, 1804, *Political Register,* Vol. V, No. 25, p. 992.

108. Lord King, 1803, quoted Horner, in Fetter, 1957, Ed., op. cit., p. 89.

109. Ibid., p. 90.

110. Idem.

111. *Edinburgh Review,* 1803, Vol. 2, No. 4, Art. xi, p. 403.

112. P. Sraffa and M. H. Dobb, 1951–55, Eds., *The Works and Correspondence of David Ricardo,* 10 Volumes (Cambridge: Cambridge University Press), Vol. X, p. 49.

113. Horner, in Fetter, 1957, Ed., op. cit., p. 77.

114. Stephen and Lee, 1917, Eds., op. cit., Vol. 11, p. 147.

115. *Edinburgh Review,* 1803, Vol. 2, No. 4, p. 408.

116. Idem.

117. *Hansard's,* Vol. XXXVI, p. 1247.

118. *Hansard's,* 1804, Vol I, p. 1084.

119. Ibid., p. 1088.

120. Ibid., pp. 1089–90.

121. Fetter, 1965a, op. cit., p. 39.

122. F. W. Fetter, 1955, Ed., *The Irish Pound 1797–1826* (Evanston, Il.: Northwestern University Press), p. 31.

123. Irish Currency Committee, 1804, *Report from the Select Committee on the Circulating Paper, the Specie and the Current Coin of Ireland* (London).

124. Irish Currency Committee, 1804, quoted Fetter, 1955, Ed., op. cit. p. 44.

125. Ibid., p. 52. It is said, nevertheless, to have formed the analytical basis for the later *Bullion Report.* Refer to Fetter, 1959, op. cit., p. 103.

126. *Political Register,* 1804, Vol. V, p. 290.

127. It has been suggested that the depreciation of Irish currency in 1803 was due in some part to the appreciation of the English pound viz a viz Hamburg following the Peace of Amiens. Refer to Fetter, 1955, Ed., op. cit., p. 25.

128. J. Wheatley, 1803, quoted Fetter, 1965a, op. cit., p. 38, and Silberling, 1924, op. cit., p. 416.

129. Silberling, 1924, op. cit., p. 417.

130. Sraffa and Dobb, 1951–55, Eds., op. cit., Vol. vii, p. 220, fn2.

131. F. W. Fetter, 1949, "The Life and Writings of John Wheatley," *Journal of Political Economy,* Vol. 50, pp. 357–76.

132. Henry Brougham (1778–1868). Lord Brougham in 1830. Born and educated in Edinburgh Brougham was called to the Scottish Bar in 1800 and the English Bar in 1808. He was a co-founder of the *Edinburgh Review* in 1802, and a key determinant of its content and direction until becoming a member of parliament in 1810. After 1815 he became a leading Whig figure.

133. *Edinburgh Review,* 1803, Vol. 3, No. 5, Art. xviii, pp. 231–252.

134. Ibid., p. 232.

135. Ibid., p. 246.

136. Ibid., p. 248.

137. Silberling, 1924, op. cit., p. 418.

138. Wheatley, 1807, quoted Silberling, 1924, op. cit., p. 419.

139. Silberling, 1924, op. cit., p. 420, fn7.

140. Fetter, 1955, op. cit., p. 55.

141. Fetter, 1955, Ed., op. cit., p. 55, and pp. 125–128.

CHAPTER 4

English Currency and the Bullion Report

*The new economic forces deployed behind the flashing swords of the dying feudal aristocracy, and drew from the financial needs and commercial opportunities of the war an additional impetus and an increased power. Capitalism joined with feudalism to fight Napoleon, and was an essential instrument in his destruction. England bought victory, as she bought her European allies, with the subsidies furnished by her money-lords. The English aristocracy won the war only by getting into debt to the English capitalists.**

Following the established pattern this chapter will begin by setting the scene for the period 1807–10, when there was a sudden renewal of controversy culminating in the production of the *Bullion Report*. After reviewing the setting we present a brief analytical preview, suggesting patterns of controversy, and of the developing modes of discourse therein. The third part of the chapter will review the arguments and theoretical contributions between 1807 and August 1810. Once again the accent will be analytical/documentary, rather than exegetical. The final part of the chapter presents an analysis of the foregoing, a brief look at the issue of the politics of the bullion report, and a glance ahead at its reception.

The outline sketch of the setting circa 1807–10 will fall into four main parts: a look at the political leadership and the war with Imperial France, a look at the war economy, a look at the financial community and its war, and a look at the social background and experiences of David Ricardo, a key player of the period.

THE SETTING: 1807–1810

The deaths of Pitt and of Fox in 1806 left both the Tory and Whig parties in disarray. They slipped back once more into the pre-party

G. D. H. Cole, 1971, *The Life of William Cobbett* (Westport, Conn.: Greenwood), p. 2.

political pattern of insular personality factions characteristic of eighteenth-century politics. The Whigs were split into Greyite[1] and Grenvilleite[2] factions, with an in-again out-again radical group led by Samuel Whitbread.[3] They were further divided after Lord Howick was elevated to the House of Lords on the death of his father in November 1807.[4] The unreconcilable Grey and Grenville wings of the party could not agree on a leader in the House of Commons save one who had no decided opinions, who offended nobody, and did nothing. George Ponsonby, Irish lawyer, was their man. His position secure after the transfer of Petty to the Lords in November 1809, Ponsonby presided over the self-destruction of the Whigs as a party until 1817.[5]

The Tories, for their part, were divided into Pittite (without Pitt), led perhaps by Lord Castlereagh more than anyone, and Canningite[6] factions. The result was that after a brief period of unity in the coalition government, the Ministry of All the Talents, an extremely weak Tory government under Spencer Perceval,[7] was allowed to stumble on until Perceval was shot by the apparently mad bankrupt John Bellingham in 1812. Added to this was the descent of King George III into fits of madness from 1810 onwards. In short, this post-1807 period witnessed a considerable increase in intra-party political factionalism, together with a marked decline in leadership. Indeed, decisive leadership was rarely more lacking in Britain's times of crisis than it was during the greater part of the wars with Imperial France.

Trade was disrupted during this period by the enforcement of Napoleon's Continental System in 1807–09, and by the blockade of European ports. The continental system was effective for a short time only, and it is usually judged to have been a failure overall. However, in looking at the trade statistics for the period it should be noted that the depreciation of English currency may well have accounted for a considerable proportion of the increase in nominal export values,[8] thus hiding the extent to which the trade disruptions of 1807–09 were felt. The disruptions to trade and Napoleon's Continental System called forth a retaliatory response from England in the form of the Orders of Council of 1808. The aim was to prevent neutrals trading with France by means of routing trade with continental Europe through English ports.

The Orders of Council were roundly criticised by nearly all leading economic figures of a liberal bent in England. They complained that it was their trade that was damaged by further restrictions, and not that of the French. This created some factional polarization, because it put the Tory ministry off-side with many merchants and bankers. Political and economic liberals began to identify with one another, and to see a common opponent in the Tory leadership. Even the more liberal

Table 4.1. Wheat Prices Per Quarter, and Bread Prices Per Loaf, 1800-1839

Years	Wheat Prices (Shillings per quarter)	Bread Prices-London (Pence per 4lb. loaf)
1800-04	84.85	11.7
1805-09	84.57	12.2
1810-14	102.45	14.6
1815-19	80.35	12.8
1820-24	57.15	10.0
1825-29	62.47	9.9
1830-34	57.67	9.4
1835-39	55.78	8.7

Source: P. Mathias, 1969, *The First Industrial Nation: An Economic History of Britain, 1700-1914* (London: Methuen and Company), p. 474.

Tories became disaffected with their party leaders, thus creating some intra-Tory divisions. It was in this environment that the *Quarterly Review* was started in 1809. While it began as a Tory party organ, and in the context of a self-appointed opposition to the *Edinburgh Review,* it soon came to espouse a rather factional Canningite line.

A poor harvest in 1809 greatly added to the economic problems of the period. It added a payments problem for the government because it created the need for corn imports. This was particularly significant because throughout the 1770 to 1850 period the corn trade was the major cause of balance of trade fluctuations.[9] And it added to the economic distress of the common people, because corn (bread) was such a major part of the wage earners' expenditure that they were especially vulnerable to corn price fluctuation. In 1809 there was a renewed marked upturn in prices (Tables 3.1 and 4.1), and public unrest soon followed. The business community also faced problems in the period. The disruption of international trade precipitated quite severe recessions in the export sector during 1807–08 and 1811–12.[10] Beginning in late 1809 there was a further sharp increase in the number of bankruptcies recorded; from 1,101 in 1808 to 1,792 in 1810. (See Table 3.3 above.)

We have noted how the wars with France had raised the stakes of the game in the City. Competition between the major City groups (merchants, bankers and Stock Exchange operatives) became increasingly open and prone to develop into outright conflict. The so-called Gentlemen of the Exchange enjoyed some success in their pursuit of status and power, by furthering functional specialization and differentiation in the financial community so as to develop the role of the Stock Market. In the years after the opening of the new Exchange in

1802 they successfully defended their subscription room status in the face of bitter opposition. They were not, however, successful in gaining government loans contracts until 1807.

On the occasion of the government loan of 1807, and of the Irish loan later that same year, a syndicate of gentlemen from the Stock Exchange, headed by David Ricardo and John Steers, was the successful contractor. The loan was distributed amongst Stock Exchange subscribers who were so delighted with the success of their challenge to the merchant and banker groups that they awarded Ricardo, Steers and Barnes each with a silver vase. Ricardo's, an object which he is said to have treasured, bore the inscription:

> Presented to David Ricardo esq. by the Subscribers to the loan of 1807 in Testimony of their unanimous approval of his conduct as joint contractor on that occasion, Whereby the just and equitable principle of mutual participation between contractor and subscriber has been so manfully asserted and so fully recognized, to the Honour of himself and his brother contractors, and to the satisfaction of the subscribers at large.[11]

Despite the elation of the Stock Exchange gentlemen on that occasion they were not successful in obtaining a government loan again until 1811. The intervening years, *mid-1808 to mid-1811,* marked, not coincidentally, the very height of the bullion controversy and of monetary debate.

Attempts to enforce the Stock Exchange's subscription room status, and thereby affect the exclusion of non-specialists, brought the leaders of the new exchange into conflict with the excluded former operatives. There was strong opposition from the Bank of England to an Exchange confined to elected membership and closed to the public. New members were ineligible if they were engaged in any other business; thus bankers and merchants were excluded. This was, of course, a divisive influence in the financial community. Indeed there was a petition to the House of Commons in 1810 calling for the opening of the Exchange to the public. Following a committee of enquiry, and the drafting and redrafting of a bill to that end, the move was finally defeated in *May 1810.* Clearly this was a manifestation of a conflict of interests and a power play between the merchant, banker and the exchange groups within the City of London, and its zenith being coextensive with the height of the bullion controversy, of monetary debate, is important to note.

Following the failure of Boyd, Benfield and Company the Goldsmid brothers, Bejamin and Abraham, took over their share of the mantle of the leadership in the City with Sir Francis Baring. The Duc de

Richelieu said: "There are six great powers in Europe: England, France, Russia, Austria, Prussia and Baring Brothers."[12] The Goldsmids became understudy to Baring Brothers, and they became very powerful. For some years the Goldsmids exercised great influence; they were closely involved with Abraham Newland, the chief cashier of the Bank of England, were the financial advisers of such aristocratic figures as Lord Yarmouth, and were known to have had the ear of Pitt.[13] The closeness of their relationship with the Exchequer is shown in the records. Between 1797 and 1810 400 million pounds of Exchequer bills were issued, 70 million pounds by the Bank of England directly and the bulk of the rest through the Goldsmids.[14] However, the stakes were high, and the risks great. These proved to be too much for Benjamin Goldsmid who hanged himself in 1808.[15]

It has been noted that the Goldsmid's "interests were . . . more outside than inside the Stock Exchange and they appear to have been far from popular with the other members."[16] It was the ability of the Goldsmids to bypass the market that made them unpopular with the Gentlemen of the Exchange, because it made them independent of the continuation facilities, which had been extremely lucrative for the jobbers, and therefore to the Stock Exchange.

In March 1810 the Exchequer announced a release of bills in an amount of eight million pounds to the public. As was common practice a large group gathered outside the Exchequer office on the appointed day some hours before the doors were opened to the public. Abraham Goldsmid, however, did not have to wait in the street. With his friend Sir John Peters, paymaster general at the Exchequer, Goldsmid entered by a private door and lodged before the issue was open to the public.[17] There followed an outcry of unfair competition and of favoritism which led to a Commons committee of inquiry, to which Goldsmid admitted making a gift of 5,000 pounds worth of stock to the clerks of the Exchequer bills office. Even with this admission no action was taken.[18]

The competition for government loans was also decried as a facade for favoritism, abuses and privileges. The Gentlemen of the Exchange campaigned for this style of dealing to end, and for genuine market competition to be allowed. Had their proposals been adopted they would have had the far from incidental effects of channelling the business through the Stock Market, and of preventing bankers and merchants bypassing the pockets of the brokers. But this short-term material interest is by no means the whole story. Our theory of sociocultural viability suggests that social relations are sustained by generating preferences that in turn reproduce those social relations.[19] The individualist will always prefer unregulated "free" competition; and the early nineteenth-century is no different in this than the late twentieth-century. What is interesting about the nineteenth-century argument is that it is not that the market optimizes allocation, the characteristic combination of individualist (market) and hierarchist (bureaucracy) modes of discourse, but rather that the market is open and fair, the characteristic combinations of individualist and egalitarian modes. Once again the point that this issue came to a head in the northern *summer of 1810,* coextensive with the height of the bullion controversy, should not be overlooked.

The loan of 1810 brought matters in the City to a head. Following something of a speculative boom over the preceding twelve months or so, the late (northern hemisphere) summer of 1810 witnessed a series of commercial failures. From July of that year *omnium* (all stocks—literally "all together") began to fall to a discount. The successful loan contractors for that year were Goldsmid and Company, Baring and Company, and Battye and Company. Their personal lives began to aggravate the decline in the funds. On 11th September Sir Francis Baring died, and doubts about the security of the loan immediately further depressed the price of funds. In the face of an increasing discount on *omnium,* and depressed since his brother's death two years earlier, Abraham Goldsmid shot himself on 28th September.[20] By then *omnium* had fallen to a discount of more than 10 per cent.[21]

During the crisis months of that summer the London newspapers regularly carried articles deploring the extent of the fall in the funds, and declaring them to be unnecessarily depressed. These articles hinted at the existence of a campaign aimed at breaking the fundholders:

> [20 September 1810]: The attempts to continue the depression of the funds are likely to be defeated.
>
> [22 September 1810]: . . . eager to convert public confusion to the promotion of their political views, and by certain jobbers, anxious to make it subservient to their pecuniary interests. The erroneous idea so industriously circulated by certain individuals that there is a depreciation of Bank currency, has undoubtedly contributed, in some degree, with the circumstance of pressure, to produce the late depreciation in the funds.[22]

Observing that the same words appeared in nearly all the newspapers, William Cobbett noted that this at least implied that the pieces were circulated, and were, therefore, themselves products of a campaign—a campaign to support the funds and to vindicate the system, a campaign of puffing.[23]

These campaigns were clearly the efforts of the opposing market and non-market factions of the City. They were a part of the struggle between them for power and status. This activity coincided with the period between the hearings of the Bullion Committee and the defeat of the *Bullion Report* in parliament. That is, exactly the period of the consideration of the *Bullion Report,* and the height of the bullion controversy.

The wars with France were called the "war of finance," but the war within the City of London was a "war in finance." It was essentially a war of attrition. The ruin of Boyd, Benfield and Company, and the deaths of Baring and the Goldsmid brothers left a relatively clear field of play for the Gentlemen of the Exchange. It certainly shook the dominance of the merchant and banker groups. The Stock Exchange syndicate led by Ricardo was, somewhat predictably, the successful loan contractor on the occasion of the next government loan in May 1811.[24]

David Ricardo (1772–1823) stands out as a key figure in the period in regard to the development of monetary theory, and of political economy generally. Quite aside from his stature as an economist, he stands out for us as the very embodiment and personification of a type. It is, therefore, interesting to make some note of his biographical details.[25]

The Ricardo family was of Spanish and Portuguese sephardic Jewish origin, but had been forced into exile by the Inquisition. By the mid-eighteenth-century the family was an established member of the Amsterdam stock exchange. Through business relating to the Seven Years War (1756–63) one of the partners moved to London. He soon established a leading place in the London Stock Exchange and Jewish communities, and sired a large family of fifteen to twenty-three children. The third born was David Ricardo. He was given a practical education to age eleven, and then spent two years at a Jewish school in Amsterdam before joining his father's stockbroking business at the age of fourteen.

When twenty-one he married the daughter of a Quaker, which not only marked his renunciation of Judaism, but also caused a rift with both families that left the couple financially estranged and alone.[26] David had made friends amongst his business colleagues and was set up in business in his own right by the banking house of Lubbocks and Forester.[27] He quickly rose to prominence in the Stock Exchange community, and within seven years was a member of the advisory committee overseeing the move to Capel Court in 1801–02. He also served on major investigative and regulative committees. One must suppose that Ricardo was greatly involved and committed to the Stock Exchange community. He was adopted and supported by them in the period of his estrangement from family and faith, his hour of need, and came to succeed handsomely at their game. Abandoned by his family, and estranged from his Jewish roots, David Ricardo turned to the stock exchange community for all his support.

Having renounced Judaism Ricardo became a Unitarian, which was perhaps the closest thing to Judaism, but which also involved him in what was then a radical dissenting sect. The radical dissension was not purely religious; there were also elements of political and epistemological radicalism.[28] The Unitarians were a center of rationalist thought, and an expression of the late eighteenth-century confrontation of rationalism—"scientific thinking"—with mysticism—mainstream religious conservatism. The great leader of the Unitarians was, of course, Doctor Joseph Priestley (1733–1804), who was also a chief inspiration of Utilitarianism.[29]

Thus we can see that a rational calculating mode of thought was doubly stamped on Ricardo's mind, through his daily Stock Exchange business and through Unitarianism. Likewise high, factional group commitments were doubly stamped, through the Stock Exchange community and through radical dissenting religious sectarianism. Ricardo was, in short, a highly committed member of excluded groups whose characteristic practices involved a rationalistic mode of thought. Ricardo's was a rather egalitarian form of individualism.

AN ANALYTICAL PREVIEW

Linking to the setting noted in the previous chapter, we recall how the war with Imperial France united Englishmen and women from without, affecting a reversal of the trend toward factional divisions, and thereby retrieving monetary debate from the brink of controversy in 1803. In setting the scene, above, it has emerged that in the period after the deaths of Pitt and Fox in 1806, the 1807 fall of the Ministry of All Talents and the return to office of the Friends of Pitt under the Duke of Portland, there was a marked increase in factionalism amongst the political leadership, together with a lack of leadership. Strong leadership and characteristic high group commitment and involvement are in many ways mutually exclusive. When nobody is really in charge or in control the group/faction becomes the more important. This is how we must see the factionalism of the Whigs and the Tories after 1807. In terms of the other dimensions, the old-style Tories still appeared to cling to the eighteenth-century conservatism that made them highly regulated. They were still in a high positive grid position. Their failing leadership, and the increasing importance of economic factors in this unprecedentedly expensive war, suggest some decline in power for the Tory elite. (See Figure 4.1.)

Amongst the economic parvenus the situation was somewhat different. As Cobbett constantly remarked, there grew up a paper aristocracy,[30] a parvenus of money lords. They can be placed high, positive on the

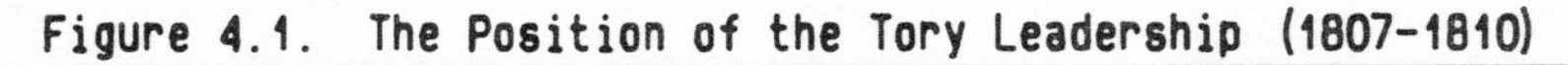
Figure 4.1. The Position of the Tory Leadership (1807-1810)

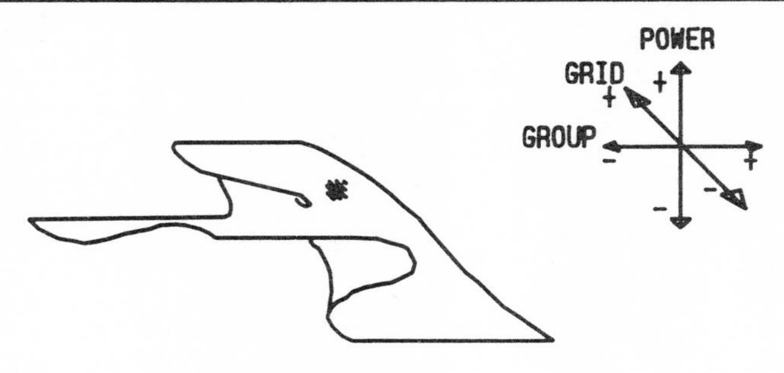

✱ Indicates Position.

Source: M. Thompson, 1982, 'A Three-Dimensional Model,' In Mary Douglas, Ed., *Essays in The Sociology of Perception* (London: Routledge & Kegan Paul), p.50.

power dimension, because of the way that the Napoleonic wars were making the money market a key focus and a key battle field. Often left to effect self-regulation in the money market and elsewhere, as we have seen, these men were clearly *not* highly regulated. They were, indeed, rule setting—low negative grid. Most marked of all was the increase in the separation of, and distinction between the functional groups in the City. In all aspects of their daily lives their experience was one of increasing factional conflict, which reached something of a zenith circa 1809–11. There was, in short, a movement toward a higher group position. Beginning from a negative group position this took them toward a zero or even positive group position. (See Figure 4.2.)

Ricardo's life and his context circa 1809–11 exemplifies the position of economic parvenu at that time. Sociologically he was ideally suited to adopting an extreme commitment and involvement in his group, the Stock Exchange. Clearly experiencing increasing wealth and power, the pursuit of status with which to compensate for his dubious cultural background (Judaism) would have been a primary goal. The chance to do this within the group, the only group that accepted him and made him its principal champion, can only have accentuated his commitment to it. He was, then, higher on the group axis than his peers. (See Figure 4.2.) This puts Ricardo in a socio-cultural position comparable to that of Walter Boyd and Lord King. And it suggests that, as the debate emerged during 1809, Ricardo would be a leading exponent of the new style of thought, of bullionism, and that he might be the chief figure in its further development.

Figure 4.2. The Position of the Economic Parvenus (1807-1810)

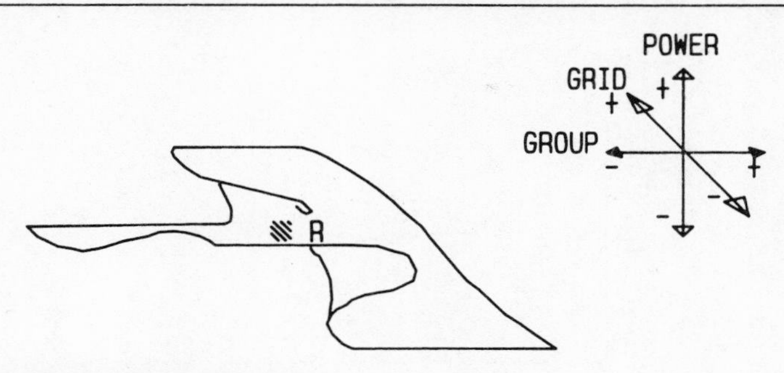

R Indicates Ricardo's position
▒ Indicates position of the economic parvenus

Source: M. Thompson, 1982, 'A Three-Dimensional Model,' In Mary Douglas, Ed., *Essays in The Sociology of Perception* (London: Routledge & Kegan Paul), p.50. Reprinted with permission.

The crucial feature of this up group movement is that it brings certain of the key players amongst the economic parvenus into the fold or cusp region in the three-dimensional model. Once within the cusp region there are available two equally plausible stable cultural biases or ways of thought, each of which matches or entails certain specific levels or measures on the three dimensions defining socio-cultural space. Importantly, the available positions match the same grid and group points with two different levels of power. (See Figure 4.3.) Positions A and B correspond to the same levels on grid and group dimensions as each other, but two different levels of power. Under such circumstances the choice between ways of life must be effected by power relations, the drive to extend and/or to maintain power. That is to say that we must expect alternative theory development and theory choice to be determined by, and within, the underlying social relations of power—specifically as these relate to the group dimension. Hence, to uncover the basis of the development of monetary thought in the context of London circa 1807–10 we must focus on the power plays of factionalized elites.

The implications for the currency debate as it emerged between 1807 and 1810 are by now obvious, and can be briefly summarized as follows. We would expect an upsurge in controversy after 1807. We would expect that there would develop an increasingly radical departure in theoretical analyses and in modes of expression. We would expect Ricardo to emerge as a leader of the new analysis and of its mode of expression.

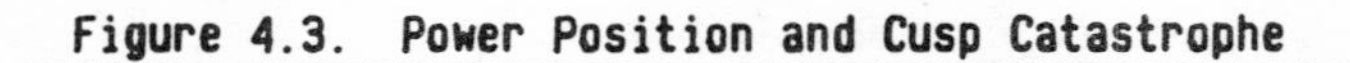
Figure 4.3. Power Position and Cusp Catastrophe

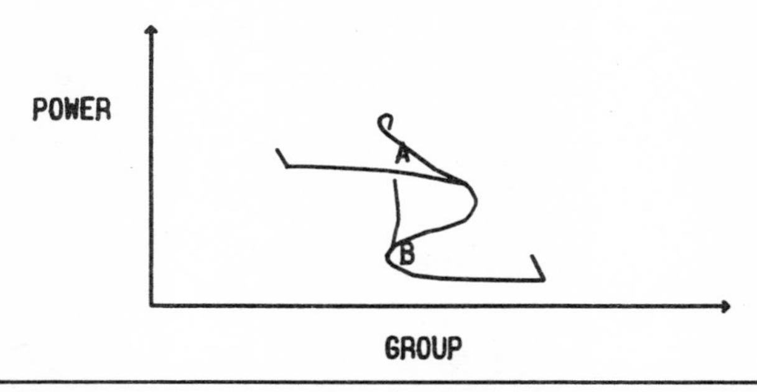

Source: M. Thompson, 1982, 'A Three-Dimensional Model,' In Mary Douglas, Ed., *Essays in The Sociology of Perception* (London: Routledge & Kegan Paul), p.49.

We might also expect a greater degree of success for the bullionist line than in 1800–07, because, as the financial needs of war grew, so the extent to which the economic parvenu could be excluded declined. Let us now turn to the monetary debate as it unfolded.

THE ENGLISH CURRENCY DEBATE: 1807–1810

After some years in abeyance the bullion controversy re-awoke during 1809. There was a rise in the premium on gold and silver to approximately 15 per cent, and a comparable depreciation in relation to foreign exchanges in 1809,[31] but we will show that this, while perhaps a factor, was *not* sufficient cause for debate. Indeed the gold price and depreciation of exchange (Hamburg) peaked some 16 and 7 percentage points higher, respectively, in 1813, but produced *no debate at that time.* (See Table 4.2.) As in the case of Irish currency, the event of currency depreciation cannot, of itself, causally explain the occurrence of debate, and thereby the development of monetary theory.

It was in 1809 that David Ricardo made his debut in print, in the *Morning Chronicle* of 29th August, with "The High Price of Gold." It was a straightforward bullionist piece in the form of an extended letter to the editor, and was followed in September and November of 1809 by further pieces in response to ensuing correspondence.[32] In "The High Price of Gold" Ricardo sought to raise the level of public con-

Table 4.2. Percentage Deviations of the Pound from Par, 1809-1824

Year	Spanish Dollars	Hamburg Exchange
1809	14.5	-19.5
1810	16.2	-18.0
1811	21.8	-30.9
1812	32.4	-24.6
1813	40.9	-26.2
1814	24.1	-12.6
1815	15.7	-10.5
1816	0.2	1.7
1817	1.5	1.1
1818	9.4	-3.7
1819	3.5	-0.8
1820	0.7	2.2
1821	1.0	3.5
1822	-1.0	2.7
1823	-1.2	3.5
1824	0.2	1.6

Source: N. J. Silberling, 1919, "The British Financial Experience, 1790-1830," *The Review of Economic Statistics* (Amsterdam: Elsevier Science Publishers), Volume 1, No. 4, p. 287.

sciousness of the importance of monetary issues, and to warn of the dire consequences of restriction. He wrote:

> The present high market price above the mint price of gold, appears to have engrossed a great portion of the attention of the public; but they do not seem to be sufficiently impressed with the importance of the subject.[33]

Ricardo argued that the increase in the market over the mint price of gold indicated an overissue of Bank of England paper, that the fall in exchanges was also indicative of this overissue, since it kept pace with the gold price changes, and that the original cause of the problems was an overissue which had been permitted by restriction vesting dangerous powers in the directors of the Bank of England.[34] Ricardo recommended an immediate phased withdrawal of paper, and the resumption of cash payments at the Bank.

Though he commenced with an argument that people demanded gold during restriction,[35] Ricardo failed to satisfactorily demonstrate that such a gold preference did not act to depreciate paper. Nor did he show that trade disruptions and remittance demands had not dislocated the equilibrating mechanism of the balance of payments sufficiently to account for the exchange variations. Indeed Ricardo did not even explain the mechanisms by which excess paper issue led to depreciation.[36] His debut was not a strong effort. It was "a reversion to the superficial pronouncements of Fox and Boyd."[37] Sir John Clapham went so far as to say that Lord King in 1804 put Ricardo's thesis of 1809 more clearly than Ricardo ever did.[38]

The great contrast between the situations of England in 1809 and Ireland in 1804, which King had failed to sufficiently consider, were completely lost on, or by Ricardo. He merely ornamented Lord King's argument with alarmist rhetoric, being anxious to deliver Englishmen from the state of things under restriction "which is pregnant with present evil and future ruin."[39] We can note that contemporaries immediately recognized Ricardo's pamphlets as being like those of Fox, Boyd and King. They were both stylistically and analytically bullionist.

More importantly we can immediately see a hint of the rhetoric which is characteristic of high group involvement and commitment—characteristic of the factionalized social environment of sectarianism. To describe the power of the Bank of England as an *evil* is typical of high group position, good within/evil outside.[40] Egalitarian relations of the high group, low grid and low power position arise when strong group boundaries are coupled with minimal prescription. The key boundary is that between the group and outside, since there are no boundaries inside the group. Such social relations generate a "wall of virtue"[41] around the group. And Ricardo saw restriction of payments at the Bank of England pregnant with present *evil.*

Late in 1809 Ricardo is reported to have undertaken a (re)read of a number of works relating to monetary theory, including those of John Locke, Sir James Steuart, Adam Smith, Lord Liverpool, and Henry Thornton.[42] He subsequently, despite Silberling,[43] re-wrote his monetary ideas in the fuller form of a pamphlet, *The High Price of Bullion; A Proof of the Depreciation of Banknotes* (London, January 1810).[44] In reformulating his views, however, Ricardo did not greatly add to his basic position or argument. On reviewing Ricardo's pamphlet in the *Quarterly Review* of February 1810, William Gifford wrote:

> He has only given a statement of those sound and orthodox doctrines respecting the effects of a superabundant paper-circulation, which are now being generally received; and his statement has the recommendation of just about as much novelty as an acknowledged doctrine may be expected to contract in passing through the thinking of an ingenious mind.[45]

Ricardo's pamphlet began from the quantity theory of money, which he attributed to Lord King, and it spent considerable effort in controverting the analysis of Henry Thornton.[46] Ricardo stated that depreciation was due to overissue and not to any lack of confidence in the Bank of England.[47] His pamphlet, letters, and his later *Principles,* however, contained hints that depreciation may have had a qualitative as well as a quantitative element, that it involved some inherent debasement of quality. Ricardo slipped easily into such language as that which follows:

> By sending 130l. good English pounds sterling to Hamburgh, even at the expense of 5l., I should be possessed of 125l.; what then could make me consent to give 130l. for a bill which would give me 100l. in Hamburgh, but that my pounds were not good pounds sterling?—they were deteriorated, were degraded in intrinsic value below the pounds sterling in Hamburgh . . . [48]

Ricardo thus re-introduced a degree of discursive equivocality on a point that Thornton's efforts had quite largely expunged. Importantly, though apparently not obviously to Ricardo, Thornton's introduction of the velocity of circulation argument had made it unnecessary for the bullionists to revert to qualitative factors in explaining deviations between the quantity and price of currency. Like other bullionists at that time, Ricardo did *not* use the available velocity of circulation argument. It is important to attempt to find out why.

It may well have been because Thornton was at that time perceived to be a defender of restriction, that is, because Thornton was outside Ricardo's group and was to be shunned—regardless of theoretical considerations—in the manner of the characteristic factional social relations and cultural biases; good inside and evil out. Indeed, of course, Thornton was associated personally with the Bank of England. Such a scenario would demonstrate how theoretical advance can be determined by sociological factors.

Ricardo's perception was that it was necessarily paper that fell in value, rather than gold or convertible foreign currency that rose. Unlike the other analysts, he gave no consideration to the demand for gold (or silver). Ricardo's interest was in maintaining the exchanges at par, and the mint equal to the market price of gold. It was not in maintaining a stable level of prices (the focus of Horner and Cobbett) or of production (the focus of Baring).[49] Ricardo's was very much a City, money market view, rather than a merchant's or producer's view. Furthermore, "as theory it was much inferior to what was in Thornton, Horner and the Irish Currency Report."[50] Ricardo's line differed from that of Thornton and Horner at both the theoretical *and* lower-case p political levels at that stage.

Ricardo's mode of analysis was also remarkable, though from the point of view of cultural theory perhaps not surprising. Mannheim identified two contributory factors in the inclination toward abstract analysis. First, those whose daily life involves them in abstract quantitative calculation naturally turn to rationalism.[51] Second, "the capacity for abstraction amongst individuals and groups grows in the measure that they are parts of heterogeneous groups and organizations."[52] The Gentlemen of the Stock Exchange were a close-knit group of outcasts; Jews, Catholics, foreigners, muckworms, . . . whose daily life involved calculation. Ricardo's abstract theoretical approach, so surprising to his contemporaries, had a clear and quite obvious social basis. There are, that is, clear sociological determinants of the methodological revolution which Ricardo affected in political economy.

In the conclusion to *The High Price of Bullion* Ricardo stressed that he was not accusing the directors of the Bank of England of interested motives, but he asserted that the consequences of their mistakes were nonetheless painful for their innocence.[53] He complained that the Bank of England lacked real independence from the government, that the government was inclined to (re)enact restriction in order to avoid the Bank recalling its huge loans. Ricardo thus began to implicitly suggest that the Tory ministry and the Bank of England were a single force, and thereby champion the cause of an independent central bank. He also spelled out the distributional effects of depreciation, and complained that fundholders were gaining, and that capitalists, the real producers of wealth, were losing. It was, he said, quoting Smith,

> enriching in most cases the idle and profuse debtor at the expense of the industrious and frugal creditor, and transporting a great part of the national capital from the hands which are likely to increase and improve it, to those which are likely to dissipate and destroy it.[54]

Ricardo thus made clear the relationship between the thrust of his analysis and the socio-political context of the day, as well as drawing the attention of the most disadvantaged section of the public to its need for concern.

Ricardo's analysis was founded on a relatively more simple quantity theory of money than that of Thornton or Horner. That much can be established, and in the context of the Bullion Committee, confirmed. This raises the key question regarding theory choice. Why adopt a new theory? Why ignore one? In the application of the theory of socio-cultural viability we are attempting to explain preference or choice by reference to cultural bias. Mainstream approaches to the question of theory choice have focused on empirically based rational choice. According to such approaches a theory or schema will be preferred if it permits more accurate predictions to be made and/or has greater empirical content—encompasses more evidence or explains more of the evidence.[55] This case, both at the level of Ricardo's choice not to use the velocity of circulation argument, and of Ricardo's ascendancy, refutes these mainstream approaches. Hence we must ask why the more simple bullionist quantity theory of money developed, and eventually triumphed in the period 1819–21.

The crucial element of Ricardo's input to the theoretical debate at this stage lay in his helping the opponents of restriction to achieve a definition of depreciation, and a measure of it. Whereas Thornton had introduced interpretive flexibility by establishing what did *not* count as evidence, Ricardo introduced a positive element into the bullionist argument by establishing what did count as an indicator, and as a measure of depreciation. Depreciation was perceived to be, and its extent perceived to be measurable by, the amount of gold price variation above and beyond the real cost of moving gold. Real cost being always positive, depreciation was thus always less than the whole amount of the exchange variation.[56] Therefore, this latter could not be the measure of depreciation; it was merely indicative. Ricardo thus offered the bullionists a path toward discursive closure.

The problem which remained at the theoretical level was the more basic one of the definition of money. Thornton's book had been, as the title suggests, about *paper credit;* while Ricardo's pamphlet was about the depreciation of *banknotes.* Clearly the latter is a much more restricted definition, and between the two lay much of the space for controversy at the theoretical level. The methodological implication is that the mainstream (positivist) notion, that definitions come from observation, is exposed as unhelpful. One cannot observe money; money is not simply an object of the empirical realm. Money is a relationship. The development of the theoretical term *money* in the emergent financial system was central to the debate—central, obviously, to the development of monetary theory and policy.

Ricardo was not the only influential pamphleteer in 1809–10. Another pamphlet that appeared just prior to the formation of the Bullion Committee was *An Enquiry into the Effects Produced on the National Currency and Rates of Exchange, By the Bank Restriction Bill,* (London 1810) by Robert Mushet. As an official at the Royal Mint Mushet was in a position to found his analysis on an extensive tabulation of data concerning rates of exchange and the market price of gold. In essence Mushet saw restriction as a serious problem, and he sought to champion convertibility. Nevertheless, he realized the difficulty of resumption under the prevailing circumstances. He concluded his pamphlet saying:

> From what has been stated in the foregoing chapters, the remedy for the evils occasioned by the Restriction-Bill must be obvious. . . . [But] [i]t is clear, that an operation of so serious a nature should be gradual.[57]

Once again the rhetorical use of the notion of *evil* is suggestive of high group commitment, an experience of factionalism.

Mushet also mentioned the fact that the Restriction Bill had afforded the Bank of England "enormous profits." Here is the clue to the background of that factionalism. As an official of the Royal Mint Mushet clearly had a vested interest in the situation. While banknotes were depreciated coins were driven out of circulation, and while specie payments were suspended coin minting was largely unnecessary. Thus any power that the issuers of coin currency, the Mint, had had was lost to the issuers of paper currency, the Bank of England.[58] The Bank of England had, under cover of restriction, been able to usurp the power and prerogative of the Royal Mint. This is the context of Mushet's bullionism, and of his early attack on Bank profiteering.[59]

During May 1810 William Blake (1774–1852) published his pamphlet *Observations on the Principles which Regulate the Course of Exchanges; and on the Present Depreciated State of the Currency* (London 1810). Blake's major contribution was to highlight the distinction between *real* and *nominal* exchange variations, which he felt had been blurred in the debate so far. He asserted that exchanges depended on the value of bills of exchange, and that this had two main determinants, the abundance or scarcity of the bills, and the relative values of the currency in which they are written and that of the place in which they appear in the market.[60] Blake maintained that the rate of exchange varied from two distinct causes, and that these must be kept distinct in analysis. He referred to the former as *real* and the latter as *nominal.* Blake also made the crucial point that with an excess issue of paper currency *both* paper and coin were depreciated.[61] The automatic remedy for depreciated coin was its melting and conversion to bullion; therefore, the disappearance of coin was the proof of its depreciation. The problem of the day, of course, was that the paper banknotes could no longer be converted into anything—they were inconvertible, always currency—and coin had already largely disappeared.

Of the *real exchange,* that relating to the demand and supply of bills, Blake demonstrated how poor harvests and subsidy payments overseas could depress the price of bills abroad, while raising their price through increased demand in the home market. Thus, for Blake, assuming that the real price of bullion was the same in all countries, the expense of transporting bullion was necessarily the limit for the variation of real exchanges.[62] Blake also argued that a continued high level of war remittances would raise the demand for bullion in the domestic market, because it was needed to send abroad to pay off the real exchange shortfalls, and so raise its market above its mint price.[63]

Of *nominal exchange* variations, those relating to the relative value of currencies, Blake stressed that they do not call for a movement of gold.[64] Nominal exchange variation, on Blake's account, arose from a change in the quantity of currency in one country with no change in the number of transactions or of the goods to be circulated.[65] Where the augmentation of a currency had altered prices and thus effected the exchanges the only remedy was to adjust the money supply. Gold would not be exported in the case of a nominal exchange variation because the profit of melting for sale at home would exceed that of transporting bullion abroad.

Blake went on to demonstrate how the exchange rate, or what he called the *computed exchange,* was made up of both real and nominal elements. Given that these bear no necessary relation to each other, the computed exchanges were seen by Blake to be a most confusing basis for analysis. He pointed out that a recent number of the *Quarterly Review*[66] contained an example of the difficulties. In the case at hand Ricardo took exception to Thornton's assignment of cause and effect in relation to exchanges and the price of gold. Thornton saw the former as the cause of the latter, while Ricardo was inclined to perceive things in the opposite way. The reviewer, probably William Greenfield,[67] added to the problem by asserting that either could be correct at various times. Blake pointed out that "[t]he apparent contradictions arise from confounding the *real,* the *nominal,* and the *computed* exchanges."[68] Thornton was correct in relation to the real exchange, Ricardo in relation to the nominal, and the reviewer in relation to the computed exchange.

Blake's extremely sound pamphlet was also interesting insofar as it too prefigured elements of a later phase of the debate. Blake outlined the relative class effects of restriction and depreciation, and noted in some detail the benefits which had accrued to the proprietors of Bank stock. He recorded that they had received a bonus of 10 per cent in 1799, 5 per cent in 1801, 2½ per cent in 1802, 5 per cent in 1804, 5 per cent in 1805, 5 per cent in 1806, and that the dividend was raised from 7 to 10 per cent in 1807. Overall the 100 pound stock had increased from 127 1/2 pounds in 1797 to 280 pounds by 1809.[69] A sleeping issue in 1810, Bank of England profits became a major part of the subsequent controversy. It is notable that one of the key features in later policy decisions entered the debate in this way, under the partisan stewardship of Robert Mushet and in the work of William Blake.

AN ANALYTICAL SUMMARY

The socio-cultural schema has alerted us to the importance of shifts along the group axis in the development of monetary theory, specifically to the approach of those of the economic parvenus with the greatest group commitments to a crucial cusp region. The passage of the debate would seem to offer considerable confirmatory support, both in the general terms of the debate and the emergence of Ricardo as a champion bullionist.

It has been shown that the event of depreciation, while a factor, was not, of itself, a sufficient cause of debate. Rather, the increasing group factionalism arising in the social environment and daily experiences of the participants can be seen to have been the chief factor. The details of timing of events correlate not depreciation and debate, but rather factional rivalry in the City of London financial community and debate. Indeed debate and factionalism appear highly correlated. The emergence of Ricardo as an exponent of bullionism is also notable. He expressed the underlying idea in an ever more simplified, abstract form, whilst the analyses of others developed in empirical complexity and subtlety. There was, in short, a development of a more extreme mode of expression amongst the bullionists just as the socio-cultural analysis suggested. The bullionist line was also clearly gaining in support. That the bullionist principles were in essence to be embodied in the recommendations of the Bullion Committee testifies to that.

The other important feature of this phase of the debate is its continued revelation of the relation between socio-cultural position and the broad perception of the key players in the debate. It is suggested herein that it is this position, rather than the analytical quality of their work, that determines their impact, their success as monetary theorists, and thereby, of course, the direction of the development of monetary theory. We refer specifically to the marginal role and recognition of William Blake, as compared with the apparently analytically inferior Ricardo with his socio-cultural schematic, discursive and sociological centrality.

Having demonstrated the coherence of the socio-cultural approach, it remains to ask how our assessment of the social environment amongst the leaders of the day helps us to produce a better history. To answer this we will examine one issue from the history of economic thought, namely, that of the politics of the bullion report.

The Politics of the Bullion Report

Growing popular and parliamentary comment on the state of the exchanges and currency led to the establishment of a House of Commons committee of enquiry, the Bullion Committee, in February 1810. On 1st February Francis Horner put forward a motion in the House of Commons for a committee of enquiry, citing two major motives: first, the real importance of the subject; and second, to clear up the misconceptions which had recently been publicly aired. He stated that for his part he did not blame the number of country banks, nor did he blame the problems wholly on overissue from the Bank of England, as had been "urged in various forms before the public."[70] Horner suggested that the causes were part overissue, and part trade disruptions and war. The motion was passed unopposed by either side of the House. It is notable that in calling for an enquiry Horner laid out the agenda in some detail, and with a good deal of insight.[71] It was later noted in the *Monthly Review* that prime minister and chancellor Perceval took no care over the selection of the members of the committee until it was too late,[72] implying surprise that the ministry did not have the foresight to stack the committee, and/or set the agenda.

Amongst the 22 members appointed to the Bullion Committee there was a fairly representative range of party affiliation. (See Table 4.3.) Included were Henry Thornton, Francis Horner, and William Huskisson; to whom the joint authorship of the *Bullion Report* is attributed. Of the report Horner wrote: "It is a motley composition by Huskisson, Thornton and myself; each having written parts which are tacked together without any care."[73] Produced in August 1810 this report formed the basis of further debate. Indeed as McCulloch suggested, "[i]t gave birth to myriads of pamphlets. But of these by far the greater number evinced the zeal rather than the knowledge of their authors."[74]

In the light of evidence presented to the committee concerning the price of gold and the state of exchanges—namely, that gold and silver were priced some 15 per cent above mint par, and exchanges had been increasingly unfavorable to England since late 1808, reaching some 16 to 20 per cent below par in the case of Hamburg in early 1810[75]—the committee suggested that the circumstances "pointed to something in the state of our own currency as the cause of both appearances."[76] In order to test this hypothesis the committee sought information on both these occurrences from witnesses drawn from the financial and mercantile communities.[77] It reported the evidence in an appendix, and outlined its analysis of the situation with reference to that evidence in the body of the report.

Regarding the price of gold, the committee recorded that in early 1810 the market price of gold stood some 15 per cent above its mint price, and found that witnesses attributed this high price to the high demand for, and scarcity of gold.[78] The witnesses suggested that this demand came from the needs of the armies, and the scarcity from the

Table 4.3. The Members of the Bullion Committee, Their Occupations, and Party Affiliation, Circa 1810

Name	Profession	Party	Vote on Report
James Abercrombie	Lawyer	Whig	for
Thomas Brand	Lawyer, Gentleman	Whig	for
William Dickinson	Lawyer	Ind.	for
Davies Giddy	Gentleman	Ind.	for
Pascoe Grenfell	Merchant	Whig	for
Francis Horner	Lawyer, Reviewer	Whig	for
William Huskisson	Lawyer, Reviewer	Tory(Canningite)	for
George Johnstone	India Board & East India	Ind.	for
Magens Dorien Magens	London Banker	Ind.	for
Henry Parnell	Irish Landlord	Whig	for
Richard Sharp	Merchant & Manufacturer	Whig	for
* Henry Thornton	London Banker	Ind.	for
* George Tierney	Gentleman	Whig	for
Alexander Baring	London Banker	Ind.	against
John Irving	London Merchant	Tory	against
* Charles Long	Paymaster General	Tory	against
* William Manning	Deputy Governor B of E.	Tory	against
Spencer Perceval	P.M. & Chancellor	Tory	against
Thomas Thompson	Merchant & Banker	Ind.	against
John Leslie Foster	Lawyer	Tory	No vote
* Richard Sheridan	Playwright	Whig	No vote
* Lord Temple	Gentleman	Tory	No vote

* Also members of the Irish Currency Committee.

Sources: J. R. McCulloch, 1966, Ed., *A Select Collection of Scarce and Valuable Tracts and Other Publications on Paper Currency and Banking* (New York: A. M. Kelley), pp. 404-405; and F. W. Fetter, 1959, "The Politics of the Bullion Report," *Economica*, Volume 26, pp. 119-120.

tendency to hoard in times of uncertainty. On this key issue the committee's analysis is somewhat clouded by its failure to distinguish between situations of convertibility and inconvertibility. Committee members suggested that a scenario such as that suggested by witnesses would imply that the price of gold would be raised relative to that of the silver-based currencies, but noted that this had not happened.[79] The committee noted that there was an unusually high level of silver importation into England,[80] and that silver was also some 15 per cent above its mint price. It did not appear to consider silver a close substitute for gold, as a hoardable commodity. It merely reaffirmed the presupposition that gold was the *conventional* and the *legal* measure of value, and stated that an increased issue of paper currency would raise prices locally unless it were convertible.[81] The committee concluded that an excess of inconvertible currency was the cause of the rise in the market over the mint price of gold.

Regarding exchanges, merchants' evidence given before the committee cited disruption of trade, especially with north Germany after Napoleon's gains there and the Berlin Decree of 1809, as the cause of exchange depreciation below par. An anonymous continental merchant, known as Mr. **** but revealed to have been Mr. Parish,[82] mentioned the following factors: difficulty and uncertainty of correspondence with England, few merchant houses operating the trade, the time involved in payment, circuitous shipping routes, large shipping costs, insurance costs, etc., and few exchange operators.[83] All factors seriously disrupted trade. The merchants generally thought that there was an excess of imports over exports,[84] but the equivocality amongst them as to a trade (volume) or payments (value) interpretation left the argument in a far from conclusive state.

Working from the assumption that under circumstances of convertibility the variation in the exchanges may not for long exceed the real cost of transporting bullion,[85] it was implied by the merchants that such disruptions as those alluded to increased that cost. Unconvinced that such disruptions were the cause of current exchange variations, the committee sought quantitative estimates of the cost of transporting gold. It found that all estimates proffered were below the extant 16 to 20 per cent exchange deviation below par.[86] Committee members, therefore, concluded that the disruptions could not be the only operative cause.[87] The committee found that the cost of transport had increased, but not by enough to account for the whole extent of the exchanges variation from par. The evidence of Sir Francis Baring, and of the anonymous continental merchant, suggested that an excess of inconvertible paper existed, and that this caused the exchanges to vary by an amount greater than the real cost of the transfer of precious metal.[88] With this opinion the committee felt inclined to concur.

The attempt at the quantification of transport costs may be seen as a move toward the reduction of the equivocality which necessarily remained in the merchants' qualitative judgements. It must, however, be noted that it was less than entirely successful. *The possibility is not the fact.* That one may estimate the cost of transporting bullion at x does not mean that any number of people could find willing carriers at that price. Indeed one might well question the acceptance of the quantitative estimates over the qualitative judgements of the operative merchants of the day. Moreover, that the committee posed the question indicates its failure to distinguish convertibility from inconvertibility. Under the extant circumstances, where little gold remained in circulation and paper was not convertible into it, the cost of its transport was a largely irrelevant question. The problem was rather that merchants could not get gold enough, could not negotiate bills except at greater than the usual discount.

Regarding bankers, the committee appeared to find the greatest difficulty in its evidence of bankers concerning the rationale of their issuing practices. None of the bankers saw any connection between either the price of gold, or the state of the exchanges, and the amount of their notes in circulation.[89] They stated clearly that they referred to neither when exercising their discretion over the accommodation of merchants (or the government). The committee argued that there was a link in both cases, and, moreover, that restriction had removed the natural operation of the relation and had thereby removed the natural means of exchange recovery, of equilibration.[90] It backed up this assertion by referring to the *Irish Currency Report* and earlier experiences in Ireland.[91] The evidence of exchange changes reacting to a sharp contraction of Bank paper in Ireland in 1754 was cited.[92] The committee pointed out that under convertibility the directors of the Bank of England had been ruled automatically by the exchange conditions and the price of gold, but that under a state of inconvertibility there was a "want of a permanent check, and a sufficient limitation of the paper currency of this country."[93]

The notion to which the bankers were wedded and which horrified the committee was the, so-called, Real Bills Doctrine. The bankers believed that overissue could not occur so long as issue was based on bills from real transactions and of a short fixed life.[94] The directors of the Bank of England were apparently convinced that there could

> be no possible excess in the issue of Bank of England paper, so long as the advances in which it is issued are made on the principles which presently guide the conduct of the Directors, that is, so long as the discount of mercantile bills is confined to paper of undoubted solidity, arising out of real commercial transactions, and payable at short and fixed periods. [These] are sound and well established principles.[95]

Following Thornton's analysis the committee insisted on the distinction between "the soundness of an individual loan and its desirability in terms of national monetary policy."[96] Committee members argued that the Real Bills criterion was a dangerously unsound policy. Thornton laid great stress on the need to distinguish between the central bank and other banks in this regard. The Real Bills criterion was adequate for general banking purposes (or under a situation of convertibility), but not for the central bank during restriction.[97] The committee accused the directors of the Bank of England of being ignorant of the implications of the situation arising from restriction, and of being unable to follow any rule from sound principle: "These opinions of the Bank must be regarded as in great measure the operative cause of the continuance of the present state of things."[98] The committee recommended that "the natural system of currency circulation,"[99] convertibility, be restored.

The Committee criticized the Bank of England for its mechanical reaction to the level of its reserves, and pointed to the important, though little-perceived distinction between internal and external drains. It suggested that the Bank should have extended credit rather than contracting it in situations, such as those in 1793 and 1797, where the drains on reserves were largely *internal.* The committee further suggested that the exchanges indicated which of the types of drain was operative.[100] At the same time the committee congratulated the government on its actions on those occasions in releasing Exchequer bills. Once more the role and the responsibilities of the Bank were at issue, and its past performance was criticized.

The opposition at the theoretical level between the bullionists and the Real Bills bankers was related to their respective structural positions. The bullionists insisted on, and presupposed, the City view of things, namely, that gold, and only gold, should be the standard and measure of the value of currencies. The directors of the Bank of England, that "Company of Merchants," as both Ricardo and Cobbett called them, took the merchants' view of things. Once again their choice of expression reveals the underlying factionalism. The banker merchants thought that any *real* goods, any commodities (not just gold, which was merely one such commodity) could be the standard of value for paper currencies. Therefore, the bankers argued, why not issue on the basis of real goods transactions?

Regarding the velocity of circulation the committee, in a section apparently penned by Thornton, sought to correct any too-mechanical a view of the relation between the quantity of money and its price. The committee stated that the number of notes could not, of itself, be a deciding point on the question of overissue. Rather "[t]he effective currency of the country depends on the quickness of circulation, and the number of exchanges performed in a given time, as well as on its numerical amount."[101] The committee further felt that the recent increase in the skill and facility of banking had been such as to reduce quantity needs,[102] specifically, if implicitly, thereby, discounting the fact of the modest quantity of issue increases from the Bank of England. (See Table 3.7 above.)

The conclusions of the committee bring us to the rather separate question of policy recommendations. The central finding of the committee was:

> That there is at present an excess in the paper circulation of this country, . . . that this excess is to be ascribed to the want of sufficient check and control in the issue of paper from the Bank of England; and originally, to the suspension of cash payments, which removed the natural and true control.[103]

The committee begged the government to stop this unnatural system; to stop its self-serving deception of the public, which it saw as based on the veiled (re)distribution effects of gain to the government as a debtor at the expense of all creditors.[104] Committee members recommended that a phased resumption of specie payments at the Banks of England and Ireland begin as soon as possible, since the alternative of legislative regulation of the money supply would be an "improper interference with the rights of commercial property."[105] In the true spirit of the Enlightenment, "natural" meant a system in which there was a harmony of private and public interests.

In concrete terms the Committee suggested that a maximum period of two years be allowed for a phased resumption, though this period might be shortened at the discretion of the Bank. During this time the equilibrium of market and mint gold prices was to be restored by the withdrawal of paper credit. The committee suggested that the Bank directors might "reduce their paper by gradual reduction of their advances to government, [rather] than by too suddenly abridging their advances to merchants."[106] The committee also suggested that a postponement of six months be granted in the event of peace coming during that period.[107]

Francis Horner submitted the report of the Bullion Committee to the House on 8th June 1810 and it was published, after some delay, in early August. Even given that the report was published too late for consideration before the summer parliamentary recess, Horner does not appear to have been in any hurry to press the matter forward. He appeared to believe that the report would lose nothing by delay.[108] He did not move the House into committee to consider the report until 6th May 1811. Horner then spent time arguing points of resolution, which, after a series of counterresolutions from Nicholas Vansittart, were all defeated along with the committee's recommendations on 15th May 1811.[109]

There was a mixture of Tories, Whigs and Independents serving on the Bullion Committee, and it has been suggested that party affiliation played little or no part in the production or approval of the report.[110] This is the conventional wisdom. Having been alerted to the centrality of group factionalism in the bullion controversy we can immediately see this notion of party affiliation to be a spurious one. Party politics only really emerged in the post-1815 period. We have noted above how the deaths of Fox and Pitt in 1806, and the lack of strong leadership, had let the two major parties slip back into the situation of divisive personal groupings and increasing factionalism.[111] In the early years of the nineteenth-century there existed series of cohorts, each gathered around a particular leading figure, which were only then loosely grouped into parties. The whole was a structure of shifting allegiances, and of mutual jealousies. Often under such circumstances one group predominated in a ministry, and were thus at that time in power. The designations ministerialists and oppositionists are sometimes used to highlight the division between the members of the ministry, and all others of any party.[112] The ministerialists, those in power, were often only a subset of a party.

What is important to note is that Spencer Perceval became prime minister in October 1809 over Canning, and that Huskisson,[113] the only Tory amongst the authors of the *Bullion Report,* was a Canningite. It was Lord Castlereagh rather that Perceval who was the real leader of the Tories at that time, and relations between him and the Canningites were nothing if not strained. Indeed, George Canning and Lord Castlereagh had fought a duel on Wimbledon Common on 21st September 1809, only shortly before Perceval's appointment to the ministry. Canning remained forever unpopular with the Old Tories.

So what was the political situation of the authors of the *Bullion Report?* Thornton was an independent who was concerned above all with the accuracy of theoretical analysis. His publication, out of kilter with the fads and demands of political debate, indicates his lack of political motivation and the direction of his real interests. Horner was a leading light amongst the new guard of reformer Whigs who were soon to be in the ascendancy. He had been amongst the leading young Whigs in Scotland, and a major force in the formation of the *Edinburgh Review,* which had become increasingly associated with the Whig party since its inception in 1802. Though not as radical as those of the Whitbread faction, which was sometimes, and sometimes not, under the Whig banner, Horner was amongst the young pro-reform group. One would suggest that he was, therefore, staunch in opposition to the old-guard Tory ministry. William Huskisson was a major figure amongst the more pro-reform group of Tories, the Canningites, who were currently out of office. None were *ministerialists.*

None of the leading figures of the Bullion Committee, the authors of the report, were in fact in any way defenders of the ruling Tory group, the ministerialists. When Fetter wrote that "Huskisson [was] a Tory associated with the Canning group of the party who had been a member of the government for several years up to 1809, but was out of office at the moment,"[114] he was exactly correct. However, he failed to really notice his own point of qualification in the sentence, and it is there that the key lies. Huskisson, Fetter's token Tory, was in fact recently ousted, and was the leading member of the pro-reform Canningites, the group opposing Pittites within the Tory party. The timing is also notable. Huskisson was member of the government *up to 1809.* It was in 1809 that the Canningite Tories founded the *Quarterly Review,* in 1809 that Canning and Castlereagh fought a dual. Tory factionalism grew to a crescendo between 1807 and 1809, as did the factional conflicts within the City of London financial community, and as, of course, did monetary debate.

Conventional historiographic approaches pay too much attention to either (empirical) events, such as depreciation, or to the rational reconstruction of the logical structure of the argument, and not enough to the complexity of the context. The conventional wisdom concerning the politics of the *Bullion Report* is but one case in point. It is suggested herein that concentration on logical form proves to be a major weakness, and that it is necessary to adopt a perspective which encourages a view of the development of ideas and arguments within the social environment of ongoing power relations. A socio-cultural analysis reveals that choices, be it of prime minister, of a mode and structure of monetary analysis or a monetary policy, are all subject to the same patterned selective process or cultural bias.

Cultural theory has aided historical accuracy and interpretation in the case of the politics of the bullion report. Socio-cultural analysis alerted us, as the story unfolded, to the importance of the group dimension, and of factionalism in the City of London. Having been thus alerted we are attuned to the ahistoricity of claims in regard to party politics. The key to historical understanding lies in cultural theory. We now turn to the period of the reception of the *Bullion Report* of August 1810.

NOTES

1. Charles Grey, Second Earl Grey 1807, (1764–1845). He was an early supporter of Fox and a leading oppositionist, a champion of parliamentary reform and of Catholic Emancipation. He was head of the government which carried the Reform Bill of 1832.

2. William Wyndham Grenville, Lord Grenville, (1759–1834). Prime Minister in the Ministry of all the Talents 1806–07. Grenville was member of parliament for Buckingham from 1782, became President of the Board of Control, and Foreign Secretary in 1791.

3. Samuel Whitbread (1758–1815). Leader of the Radical reformists amongst the Whigs until they were forced out of the party in 1809. He was the heir to the London Brewing business and member of parliament for Bedford from 1790. He was always a vocal opponent of Pitt and later of Dundas (Lord Melville). His calls for peace at all costs from 1807 caused a rift between he and the Whig leaders and led to the effective disbanding of the party in 1809. In later life Whitbread was a major patron of the Dury Lane Theatre. He committed suicide on the 6th. July 1815.

4. M. Roberts, 1935b, "The Leadership of the Whig Party in the House of Commons 1807 to 1815," *English Historical Review,* Volume 50, p. 620.

5. Ibid., p. 635.

6. George Canning (1770–1827). He became famous in literary circles through the *Anti-Jacobin Review* and later the *Quarterly Review,* and politically as Foreign Secretary in the Ministry of all the Talents. He was, with Huskisson, a supporter of free-trade when in office in 1823–26, toward the end of which period Canning was Prime Minister.

7. Spencer Perceval (1762–1812). 'Little P'. He was Prime Minister from 1809 until his assassination in 1812. Became member of parliament for Northampton in 1796, and was a supporter of Pitt in parliament. Became Chancellor of the Exchequer in March 1807, and Prime Minister after the death of the Duke of Portland. His support of the Orders of Council for the European Blockade was always his weak point, since many saw the Orders as more damaging to British commerce than to the French.

8. P. Sraffa and M. H. Dobb, 1951–55, Eds., *The Works and Correspondence of David Ricardo,* 10 Vols. (Cambridge: Cambridge University Press), Volume III, p. 141.

9. P. Mathias, 1969, *The First Industrial Nation: An Economic History of Britain 1700–1914* (London: Methuen), p. 232.

10. J. E. Cookson, 1982, *The Friends of Peace; Anti-War Liberalism in England 1793–1815* (Cambridge: Cambridge University Press), p. 215.

11. E. V. Morgan and W. A. Thomas, 1962, *The Stock Exchange: Its History and Functions* (London: Elek Books), p. 50.

12. Quoted B. Gordon, 1976, *Political Economy in Parliament, 1819–1823* (London: MacMillan), p. 195.

13. S. R. Cope, 1942, "The Goldsmids and the Development of the London Money Market during the Napoleonic Wars," *Economica,* Volume 9, pp. 185–206.

14. Ibid., p. 188.

15. D. Weatherall, 1976, *David Ricardo; A Biography* (The Hague: Martinus Nijhoff), p. 67.

16. Morgan and Thomas, 1962, op. cit., p. 49.

17. Cope, 1942, op. cit., p. 189, and Cobbett, *Political Register,* Vol. 18, No. 16, pp. 520–21.

18. Cope, 1942, op. cit., p. 189.
19. M. Thompson, R. Ellis, and A. Wildavsky, 1990, *Cultural Theory* (Boulder CO: Westview Press), p. 66.
20. F. W. Fetter, 1965a, *The Development of British Monetary Orthodoxy 1797–1875* (Cambridge, Mass: Harvard University Press), p. 60; P. Sraffa and M. H. Dobb, 1951–55, Eds., op. cit., Volume X, p. 81.; S. R. Cope, 1942, op. cit., p. 205.
21. Sraffa and Dobb, 1951–55, Eds., op. cit., Vol. X, pp. 80–81.
22. Quoted by William Cobbett, *Political Register,* Vol. 18, No. 16, p. 525.
23. Ibid., p. 524.
24. Sraffa and Dobb, 1951–55, Eds., op. cit., Vol. X, pp. 80–81.
25. Refer to Sraffa and Dobb, 1951–55, Eds., op. cit., Vol. X.; J. R. McCulloch, 1888, *The Works of David Ricardo* (London: Murray), pp. xv–xxxiii; and Weatherall, 1976, op. cit., for detailed biographic studies.
26. It is interesting to note that in his father's will all the children, male and female, inherited equally—except David; who was merely paid a token for acting as executor. This was most unusual at that time, and suggestive of an egalitarian background. David Ricardo did not follow this example in his own will, when the time came.
27. Sraffa and Dobb, 1951–55, Eds., op. cit., Vol. X, p. 68.
28. Weatherall, 1976, op. cit., pp. 61–66.
29. L. Stephen, 1900, *The English Utilitarians,* 3 Vols. (London: Duckworth), Volume I, and E. Halévy, 1952, *The Growth of Philosophical Radicalism* (London: Faber and Faber).
30. William Cobbett, 1804, *Political Register,* Vol. V, No. XXIV, etc.
31. Fetter, 1965a, op. cit., p. 39.
32. Sraffa and Dobb, 1951–55, op. cit., Vol. III, pp. v–vi. It is worthy of note that the *Morning Chronicle* was a Whig newspaper, and that the proprietor, Mr Perry, was a friend of Ricardo's.
33. Ibid., p. 15.
34. Idem.
35. Ibid., p. 17, and p. 22.
36. *Quarterly Review,* 1810, Vol. 3, No. 5, Art. xii, p. 157.
37. N. J. Silberling, 1924, "The Financial and Monetary Policy of Great Britain during the Napoleonic Wars," *Quarterly Journal of Economics,* Volume 38, 2 parts, p. 426.
38. J. Clapham, 1977, *The Bank of England. A History,* 2 Vols. (Cambridge: Cambridge University Press), p. 22.
39. Sraffa and Dobb, 1951–55, op. cit., Vol. III, p. 15.
40. Refer D. E. Owen, 1982, "Spectral Evidence: The Witchcraft Cosmology of Salem Village in 1692," in Mary Douglas, Ed., *Essays in the Sociology of Perception* (London: Routledge and Kegan Paul), p. 278.
41. Thompson, Ellis and Wildavsky, 1990, op. cit., p. 9.
42. Sraffa and Dobb, 1951–55, Vol. III, p. 7.
43. Silberling, 1924, op. cit., p. 423, fn2.
44. Sraffa and Dobb, 1951–55, op. cit., Vol. III, pp. 5–9.

45. *Quarterly Review,* 1810, Vol. 3, No. 5, Art. xii, p. 156.

46. Refer; C. F. Peake, 1978, "Henry Thornton and the Development of Ricardo's Economic Thought," *History of Political Economy,* Volume 10, No. 2, pp. 193–212; D. A. Reisman, 1971, "Henry Thornton and Classical Monetary Economics," *Oxford Economic Papers,* Volume 23, pp. 70–89; M. Perlman, 1986, "The Bullion Controversy Revisited," *Journal of Political Economy,* Vol. 94, No. 4, pp. 745–62; H. G. Grubel, 1961, "Ricardo and Thornton on the Transfer Mechanism," *Quarterly Journal of Economics,* Vol. 75, pp. 292–301.

47. Sraffa and Dobb, 1951–55, op. cit., Vol. III, p. 51.

48. Sraffa and Dobb, 1951–55, op. cit., Vol. I, p. 148.

49. Silberling, 1924, op. cit., p. 425.

50. F. W. Fetter, 1959, "The Politics of the Bullion Report," *Economica,* Vol. 26, p. 104.

51. K. Mannheim, 1953, "Essays in the Sociology of Knowledge" in P. Kecskemeti, Ed., *Essays on Sociology and Social Psychology, By Karl Mannheim* (London: Routledge and Kegan Paul), pp. 85–87.

52. K. Mannheim, 1936, *Ideology and Utopia; An Introduction to the Sociology of Knowledge* (London: Routledge and Kegan Paul), p. 272.

53. Sraffa and Dobb, 1951–55, op. cit., Vol. III, p. 95.

54. David Ricardo, quoted Sraffa and Dobb, 1951–55, op. cit., Vol. III, pp. 97–98.

55. The philosophical models derived from Popper and Lakatos respectively.

56. R. S. Sayers, 1953, "Ricardo's Views on Monetary Questions," in A. W. Coats, 1971, Ed., *The Classical Economists and Economic Policy* (London: Methuen), p. 36.

57. R. Mushet, 1810, quoted *Monthly Review,* Vol. 63, p. 178.

58. F. W. Fetter, 1957, Ed., *The Economic Writings of Francis Horner in the Edinburgh Review 1802–06* (London: London University Press), p. 52.

59. It was in this way that Bank of England profiteering entered the debate as an element of the bullionist argument.

60. W. Blake, 1810, in McCulloch, 1966, Ed., op. cit., p. 480.

61. Ibid., p. 522.

62. Ibid., p. 491.

63. Ibid., pp. 492–93.

64. Ibid., p. 509.

65. Ibid., p. 500.

66. *Quarterly Review,* 1810, Vol. III, No. V, Art. XII.

67. F. W. Fetter, 1958, "The Economic Articles in the Quarterly Review and their Authors, 1809–52," *Journal of Political Economy,* Vol. 66, 2 parts, p. 156.

68. Blake, 1810, op. cit., p. 534.

69. Ibid., p. 544.

70. *Hansard's,* 1810, Vol. XV, p. 270.

71. Ibid., pp. 270–72.

72. *Monthly Review,* 1810, Vol. 63, p. 284.

73. Horner, in a letter dated 26th. June, 1810, quoted by E. Cannan, 1925, *The Paper Pound of 1797–1821* (London: P. S. King and Son), p. xxiii.

74. McCulloch, 1966, Ed., op. cit., p. vii.
75. Bullion Committee, in McCulloch, 1966, Ed., op. cit., pp. 406–07.
76. Ibid., p. 407.
77. *Not* the Stock Exchange community, note.
78. Bullion Committee, in McCulloch, 1966, Ed., op. cit., pp. 406–408.
79. Ibid., p. 409.
80. Ibid., p. 412.
81. Ibid., p. 418.
82. Sraffa and Dobb, 1951–55, Eds., op. cit., Vol. III, pp. 427–34.
83. Bullion Committee, op. cit., pp. 421–23.
84. Ibid., p. 420.
85. Ibid., p. 426.
86. Ibid., p. 428.
87. Ibid., pp. 432–34.
88. Ibid., p. 421.
89. Ibid., p. 435–36.
90. Ibid., p. 437.
91. Ibid., p. 441.
92. Ibid., p. 443.
93. Ibid., p. 444.
94. Idem. See also M. Perlman, 1989, "Adam Smith and the Paternity of the Real Bills Doctrine," *History of Political Economy,* Vol. 21, No. 1, pp. 77–90, and W. Santiago-Valiente, 1988, "Historical Background of the Classical Monetary Theory and the 'Real-Bills' Banking Tradition," *History of Political Economy,* Vol. 20, No. 1, pp. 43–63. The Real Bills Doctrine later became known as the 'Law of Reflux'.
95. Ibid., p. 448.
96. Fetter, 1965a, op. cit., p. 42.
97. Ibid., p. 41.
98. Bullion Committee, op. cit., p. 456.
99. Ibid., p. 455.
100. Ibid., p. 461–62.
101. Ibid., p. 459.
102. Ibid., p. 460.
103. Ibid., p. 468.
104. Ibid., pp. 469–70.
105. Ibid., p. 470.
106. Ibid., p. 463.
107. Ibid., p. 472.
108. Cannan, 1925, op. cit., p. xxv.
109. Ibid., p. xxvi.
110. Fetter, 1959, op. cit., pp. 106–07.
111. Roberts, 1935b, op. cit., p. 620.
112. Weatherall, 1976, op. cit., pp. 137–38.
113. William Huskisson (1770–1830). Secretary of Treasury 1804–05, and 1807–09. Came into cabinet under Lord Liverpool in 1814.
114. Fetter, 1959, op. cit., p. 106.

CHAPTER 5

The Failure of Bullionist Monetary Theory and Policy

*A theory . . . is wrong if in a given practical situation it uses concepts and categories which, if taken seriously, would prevent man from adjusting himself at that historical stage.**

This chapter concerns the brief but crucial period between the presentation of the *Bullion Report* in August 1810 and the defeat of its analysis *and* recommendations in parliament in May 1811. While it is a relatively short period it is vital to an understanding of the nature of the controversy. We will recap some points regarding the social setting of the participants and the analysis of that setting, and then turn immediately to the unfolding debate.

THE SETTING: 1810–1811

It has been suggested in previous chapters that there was a sudden upsurge in factional conflict amongst the participants in the debate—both political and economic leaders—in 1808–10, and this has been related to the reemergence of the bullion controversy. To find out why the analysis and recommendations—the theory and the policy—of the Bullion Committee failed to gain sway in 1811 it is necessary to set the debate against that background.

It was revealed in the last chapter that the Gentlemen of the Exchange had engaged in the pursuit of status, the chief means at their disposal being functional specialization and differentiation of the operative groups in the City financial community. These group dynamics,

---

K. Mannheim, quoted, P. Hamilton, 1974, *Knowledge and Social Structure; An Introduction to the Classical Argument in the Sociology of Knowledge* (London: Routledge and Kegan Paul), p. 103.

together with the increase in competition and risk, with the enormous increase in the stakes of the game of war finance, lay at the hub of the changing social environment, and thereby at the hub of the development and application of monetary theory. These joint forces had led to increasingly hostile conflict within the financial community, conflict which reached a zenith during 1810. The Exchequer bills release of March 1810 embroiled Abraham Goldsmid in a dispute over favoritism, and led to a parliamentary inquiry in the northern hemisphere summer of 1810. The loan of 1810 fell to a discount during that summer, and autumn brought the deaths of the chief loan contractors, Sir Francis Baring and Abraham Goldsmid. This was the time of the publication of the *Bullion Report,* and the context of its reception.

The deaths of these leading figures produced a *de facto* change in the balance of power amongst the City factions. Through sheer attrition the Gentlemen of the Exchange gained some ground, and there is nothing like winning games to foster team spirit. Stock Exchange group involvement and commitment must be expected to have redoubled. At the same time the demands of war finance were making it harder for the political elite to exclude the economic parvenu.

In the face of this it seems that the opposing group(s) rallied in defense of their ideas, and their position. A group from the Bank of England and from government got together to launch an attack on the *Bullion Report* during late August 1810.[1] There was a rather odd behind-the-scenes play involving four shadow actors: Sir Francis d'Ivernois, a former Genevese refugee, British secret agent and close friend of Nicholas Vansittart; John Herries (1778–1855), secretary to Vansittart, financial adviser to Tory leaders, commissionary-in-chief, a friend of Nathan Rothschild, and a lifelong opponent of Huskisson;[2] Henry Beeke, an Oxford professor, backroom Tory adviser, and a friend of Vansittart; and Jasper Atkinson, something of a mystery person, though a prolific pamphleteer of that period.[3]

During September 1810 these men were engaged in preparing an opposition to the *Bullion Report.* There emerged an argument, revealing the stamp of d'Ivernois's input, to the effect that the *Bullion Report* was damaging to Britain because it telegraphed to the continent and to America that Britain was in a position of economic crisis, and was soon to be thrown into a sharp deflation. The French and the Americans were said to be delighted with the news, and looking forward to the adoption of the committee's recommendations. This line of argument emerged in subsequent debate amongst the Castlereagh-Vansittart group of the Tory ministerialists and opponents of resumption.[4]

AN ANALYTICAL PREVIEW

The foregoing suggests that the economic parvenus continued their ascendance, and, importantly, continued to experience increasing group involvement and commitment. They were moving toward a higher group dimension position. In the face of this it emerges that the Tory elite, the old elite, were becoming conscious of the assault on their position. This fostered a regrouping and a resurgence of commitment on their part; a recommitment to their view, to their ways, and to their values. They too were moving higher along the group dimension. This would suggest that some clarification of positions might be expected, a clarification of the contrary positions in the mode and content of monetary analysis and policy prescriptions circa 1810–11. It also suggests that clarification of position could be most expected amongst the defenders of restriction, the Tory and Bank group(s), with something more akin to mere restatement of position from the economic parvenus, and a conversion of more of them to Ricardo's previously too-extreme bullionism.

The City factionalism went hand in hand with the continued up group movement of the economic parvenus. There is something vitally important to notice about this analysis and the position in the socio-cultural schema. Beyond the fold region a new position, higher along the group dimension, must necessarily lead to a rapid decline along the power dimension. In theory this presents the economic parvenus with a dilemma. They must either pull back from further factionalism, or lose some of their power. (See Figure 5.1.) This suggests that we might expect a sudden cessation of controversy, and some reconciliation amongst the financial community groups. This would be necessary if they were not to lose their power. That is, effectively, a stepping back from the brink of the fold.

Our theory of socio-cultural viability, then, would suggest that reaching a crescendo in 1810, for all the reasons outlined in the last chapter, controversy would see a restatement of the bullionist position and further recruitment and persuasion to it, a hardening and extension of the anti-bullionist position, and a sudden cessation of controversy and a newfound unity among the City groups.

DEBATE ON THE BULLION REPORT

The course of debate after the publication of the *Bullion Report* falls into two main parts: first, the frantic campaigns between publication of the report and its defeat in parliament, between August 1810 to May

Figure 5.1. The Position of the Economic Parvenus Circa 1810

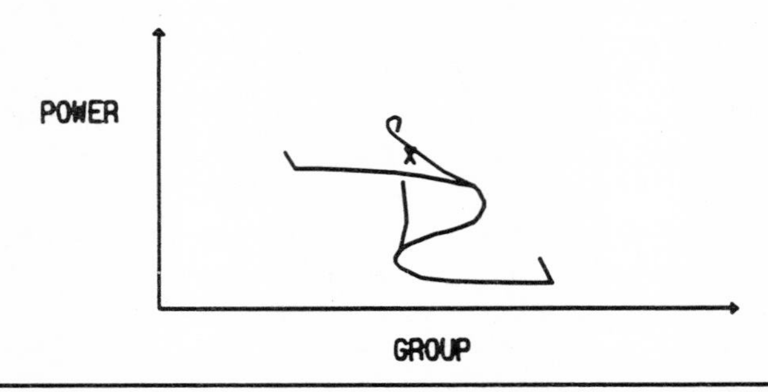

Source: M. Thompson, 1982, 'A Three-Dimensional Model,' In Mary Douglas, Ed., *Essays in The Sociology of Perception* (London: Routledge & Kegan Paul), p.49.

1811; and second, the debate which continued after the defeat of the report.

In early 1811 the orchestrations of the four backroom players resulted in the publication of the pamphlet *A Review of the Controversy Respecting the High Price of Bullion, and the State of our Currency* (anonymously) by John Herries. Herries wrote on behalf, if not at the behest, of practical men—men, that is, whose style of thought embraced the notion of empirical complexity: conservatives, hierarchists. Similar points were made by John Hill in *An Inquiry into the Causes of The Present High Price of Gold Bullion in England, and its connection with the State of Foreign Exchanges* (1810). Hill's pamphlet took the form of a long series of letters, in which he argued that the scarcity of bullion, rather than an excess issue of banknotes, was the problem. He concluded by saying that "the *desirableness* of a return to the system of cash payments [is] a very different thing from its *practicability*."[5] In *Observations on the Present State of the Currency of England* (1811) by the Earl of Rosse, formerly Sir Laurence Parsons, the *Bullion Report* was more directly challenged. Lord Rosse asserted that the price of gold was neither a sign nor a measure of depreciation, and he warned of the dire consequences of resumption at that time. He suggested that since the problems arose from trade disruptions, people should look to trade and not to the money system for an answer.[6] True to the group's overall plan Lord Rosse concluded that the *Bullion Report* produced "a triumph in the mind of the Gallic Ruler."[7]

Meanwhile, in parliament, Lord Castlereagh championed restriction as a means of isolating Britain's monetary system from Europe. He said:

> How absurd it would be for us to suffer our immense transactions at home to be deranged, by attempting to conform them to all the violent fluctuations which the enemy's lawless power can give to the continental exchanges, and through them, to the price of bullion.[8]

Once again the orchestrated opposition argument was uppermost.

Independently, as was his wont, Sir John Sinclair, president of the Board of Agriculture and doyen of enclosure and model farming,[9] offered Prime Minister Perceval his services as an opponent of the *Bullion Report* and champion of paper credit. In early September 1810 Sinclair published the pamphlet *Observations on the Report of the Bullion Committee* (London 1810), and sent a copy of it to every member of parliament.[10] He followed it up shortly afterwards with a little changed restatement. Sinclair's pamphlets came from the heart of Tory opposition to the recommendations of the Bullion Committee, in more senses than one. He declared that he was

> convinced indeed that the prosperity of the country . . . , nay, that the very safety and existence of the British Empire depended on the preservation of our *present system of circulation.*[11]

Sinclair based his analysis on the assertion that the exchange system had broken down for want of middlemen, an element often mentioned by the merchants who gave evidence before the Bullion Committee.

When reviewed in the Canningite *Quarterly Review,*[12] Sinclair was treated quite roughly. The reviewers wondered what had induced or qualified "Sir John to decide on all matters of finance," about "the anxiety of the Mercantile Body to engage him as their principal and favorite champion," and about the "zeal with which he undertook their cause."[13] More substantially the reviewers pointed out that Sinclair had performed a complete *volte-face* since suspension in 1797, when, they recalled, he wanted to "maintain convertibility." The *Quarterly* reviewers hinted at interested motives.

The *Monthly Review* writers were considerably more specific. They hastened to make "notice of a most singular circumstance in the literary career of Sir John Sinclair—no other than that he, who is now a strenuous advocate of the continuance of the Suspension-Act, was formerly one of its most ardent opponents."[14] The reviewers suggested that the change proceeded "from a singular affection on the part of the worthy Baronet for the interests of agriculture."[15] In support of their point they quoted Sinclair himself, who had suggested that, with the limitation on accommodation at the banks which a contraction of paper would entail, "the landed and farming interests would suffer perhaps in a still greater degree. They are at present enabled to go [on], . . . in consequence of the *additional prices* which their commodities fetch."[16] The most specific accusation of interested motives came from Samuel Whitbread, who noted in parliament that Sinclair had changed his mind on currency "just in time to be made a Privy Counsellor."[17]

What also stands out is Sir John's revealing choice of words. In 1797 he sought to *maintain convertibility,* while in 1810 he sought the *preservation of our present system of circulation.* In his speech to the House of Commons in defense of the Vansittart-led opposition to the *Bullion Report* on 15th May 1811 Sinclair continued to argue that, on "[t]he maintenance of the [inconvertible] currency . . . must depend the future prosperity, or the entire ruin, of this great and powerful empire."[18] A notable consistency lies in his opposition to change. Contradictory in substance, Sinclair's writings display a consistent intent, a perfect example of reactionary Toryism. On speaking against the *Bullion Report* in parliament Sinclair was reported to have declared that he wished to preserve England's "established religion, its established government, and its established currency,"[19] the link between the two latter being a quite real one.

While Sinclair was not involved in the local City factionalism (being principally a Scottish landowner/farmer) his analysis could be used in support of restriction. Like that of many older-style Tories, Sinclair's was a high grid social environment, where to live according to the current rules and to resist any change to them is paramount. Thus, while only peripherally involved, Sinclair displayed the Tory style of thought. Indeed it was this, rather than content, that was consistent in his work.[20]

Others close to the Bank of England and/or the ministry began to publish works in opposition to the *Bullion Report.* These included Charles Bosanquet, the son of a former governor of the Bank of England, and Randle Jackson, widely held to be one of the Bank's hired prize fighters.[21] As a Bank of England stock-holder, Ricardo was present on 20th September 1810 when Randle Jackson[22] addressed the general court of the Bank. The speech was reported at length in the *Morning Chronicle* the next day, and was published as a pamphlet soon afterwards. Jackson argued the predictable line that the *Bullion Report* had damaged commercial confidence, and that it and its adoption were more in the interest of Britain's enemies and trade competitors than of

Britain. He also argued that the report went against the evidence presented to the committee. Above all Jackson inferred that the *Bullion Report* was inspired by party motives.

On 24th September the *Morning Chronicle* published a letter by David Ricardo in response. Ricardo stated that he had hoped to hear a scientific discussion of the issues, but had instead been suffered to hear an attack on the party spirit of the *Bullion Report.*[23] The point that Jackson raised as to party spirit, and the tone of the series of letters from Ricardo, reveals documentary evidence of the centrality of City factionalism in the development of monetary theory. Ricardo's axis always manifested as a City one. He attacked the directors of the Bank of England as an interested "company of merchants," and more commonly later, as "practical men" ignorant of the principles of political economy. In response to Randle Jackson's comments on party spirit Ricardo referred to the "directors and their defenders."[24] The thrust of many writers, especially the Tory defenders of restriction was, rather, wartime needs—the defense of King and Country argument. Yet, even in this case, when accused of Whig party inspiration Ricardo did not reference the wider political argument.

This has since lent Ricardo's writing an aura of scientific disinterestedness, but this is misleading. It depends on just the kind of historical insensitivity noted in the context of the Bullion Committee in the previous chapter. Having been alerted to the centrality of group factionalism by means of socio-cultural analysis we can see that this appearance of disinterestedness has arisen because, hampered by inadequate historiographic methods, Ricardo's City-oriented plays were, and have remained, transparent to many not involved in the City factions. Ricardo saw where Jackson was coming from, and in his response he left the documentary evidence of the centrality of the City factionalism in the development of monetary theory and policy. Jackson was coming from the Bank of England, and Ricardo from the Stock Exchange. This is the crux of their partisanship, not embryonic political party alignment as such. It is also, of course, at the foundation of their styles of thought and the cultural bias in their perceptions—Ricardo the individualist entrepreneur with high group commitment, and Jackson the bureaucrat and spokesman for the hierarchy.

In reply to Jackson's specific point, that the committee concluded in contradiction of the evidence of the witnesses, Ricardo pointed out that the full evidence was in fact mixed, such that any conclusion could have been said to have been contrary to some of it.[25] In any event, he argued, the point was to gain information from practical men as the basis for theorizing by theorists—political economists. Ricardo argued, by way of analogy, that "[g]lass-makers and dyers are not necessarily chemists, because the principles of chemistry are intimately connected with their trades."[26] Nevertheless, the issue of the evidence surfaced again during parliamentary debate, and was, after September 1810, a point of some interest.

The fact was that the evidence in line with the committee's conclusions was derived from the then-dead Sir Francis Baring and the anonymous continental merchant, both beyond the reach of further interrogation. While it is true, as P. R. Hoare[27] pointed out, that "the contradiction regards not the testimony but the speculative opinions of the witnesses,"[28] the fact remained that all those witnesses still talking were, basically, opponents of resumption. Though an entirely chance situation, this clearly undermined the pro-resumption case at that time.

In late November 1810 Charles Bosanquet[29] published his *Practical Observations on the Bullion Report.* He put forward a painstaking critique of the report, arguing that it did not square with the facts. Bosanquet's was "regarded at the time as the most effective of the criticisms published."[30] Ellis and Canning, writing in the *Quarterly Review,* declared:

> Mr. Bosanquet presents himself as one of the most formidable champions against the Bullion Committee, and professes to fight them not with arguments but with *facts.*[31]

Bosanquet wrote of the report: "Strongly as reason appears to sanction it, [it] is at variance with fact."[32] He was the first opponent of the Bullion Report to transcend the rather barren argument that the committee had failed to pay sufficient attention to the evidence of witnesses.

Anyone could have responded to Bosanquet's pamphlet, but as our analysis might suggest it was Ricardo, closest amongst his peers to the fold (cusp) and most involved and committed in the factional rivalries, who did so. Ricardo's *Reply to Mr. Bosanquet's Practical Observations on the Bullion Report* (London 1811), written during December 1810 and published early in January 1811, took the form of a defense of the report, which, Ricardo declared, gave the "just *principles* which should regulate the currency of nations."[33] We are no longer surprised that one protagonist wielded facts and the other principles.

Bosanquet began on the topic of exchanges, using the tables of Robert Mushet to show that there had *in fact* been exceptions to the rule (that variations in the exchanges can never for long exceed the expense of transporting bullion) which the Bullion Committee put forward. Of Bosanquet's attack two things could have been said, but Ricardo confined himself to one—a matter of fact. In the second edition of his work Mushet admitted a radical error in his calculation of par, which, as Ricardo pointed out, once corrected undercut Bosanquet's case completely.[34]

The other notable point, which the *Monthly Review* mentioned, but which Ricardo did not note, was that the rule applied to a situation of convertibility, and from the statements of the Bullion Committee it followed "as a kind of corollary, that in the case of a non-convertible currency, the deviation from par may continue for an indefinite time *above* that limit."[35] That neither Bosanquet nor Ricardo noticed this point is indicative of the continued failure of the analysts of the era to clearly distinguish between situations of convertibility and inconvertibility. It also indicates the centrality of styles of thought, because Bosanquet and Ricardo argued in terms of facts versus principles, empirical complexity versus a reduction to simple causal relations, overlooking all else—blinded by their respective cultural bias or cultural filter to all else.

One of Bosanquet's other main points was much more telling, and as Ricardo's reply continued it was he, and not Bosanquet, who began to falter. Bosanquet pointed out an error in the calculations of the Bullion Committee in relation to the balance of payments, and he showed that payments between England and the European continent had in fact recently been unfavorable.[36] It was, once again, a point of detail, not one of principle.

It was in the area of trade that the opponents of the *Bullion Report* scored most of their points, and there that they left Ricardo, the money man, patently out of his depth. The *Monthly Review* expressed what many analysts felt when, of Ricardo's *Reply to Bosanquet,* it wrote:

> He attempted no comparison between the present state of our trade, and that which preceded the irregularities in our money-system. He takes no notice of our annual loss of remittances, to the extent of four or five millions, by the stoppage of the American trade to the continent of Europe; nor of the heavy drains from us for corn in consequence of the deficient harvest of 1809; nor, finally, of the impossibility, under the present circumstances, of our balancing these disadvantages by countervailing exports to the continent. These are the events which, operating on a *non-convertible paper-currency,* appear to us to have produced our present unfortunate position.[37]

Ricardo went further than the Bullion Committee in "inattention to the influence of the course of trade on exchange."[38] He ignored all but the monetary causes. This was, of course, a consequence of a style of thought centerd on simple principles, rather than empirical complexity. Cultural bias and context can be thought of as opposite sides of the same coin. This inattention to trade factors went hand in glove with apportioning blame and/or repossessing power from the Bank of England. Had Ricardo allowed trade causes, the Bank could not have been held responsible for depreciation.

Summarizing the Bosanquet-Ricardo exchange the *Monthly Review* wrote: "Mr. Ricardo's outset is promising, and his success in the early chapters appears complete: but as he advances he allows himself, perhaps, to fall too much into the spirit of the argument."[39] Indeed Ricardo's reply grew less convincing by the page. The reverse was the case of Bosanquet's effort, and he "contributed to shake the credit of the Report of the Committee among men of business."[40] Among such practical men, and many others, a major fault of Ricardo's response was his inclination to argue hypothetically, abstractly. On the question of excess issue, for example, he argued an hypothetical case to demonstrate the principal point. That is to say that Ricardo addressed the question of what would happen if excess issue existed, with the aim of saying, voilà! Most analysts at that time, and certainly the men of business, sought on the other hand to address the more practical question: Does an excess actually exist?[41]

Of the exchange between Bosanquet and Ricardo, Ellis and Canning wrote in the *Quarterly Review* of February 1811:

> Mr. Ricardo has fortunately delivered us from the necessity of endeavouring to reconcile positive facts and admitted impossibilities.[42]

Indeed the crucial role of Ricardo's interjections at that point lay in their marked tendency to reduce discursive equivocality, and to thus offer closure. From all that has been said above it is clear that it was this, and *not theoretical nicety* or greater empirical content, that Ricardo had to offer.

While all this was going on William Huskisson wrote *The Question Concerning the Depreciation of our Currency Stated and Examined,*[43] which was published in October 1810, but which had been written before (around August 1810) without the benefit of the opposition publications of Sinclair and Jackson.[44] It is said to have been one of the most widely circulated of the bullion controversy pamphlets, seeing some eight editions by 1819.[45] During the last months of 1810 and into 1811 Huskisson, Canning, Ellis and Gifford orchestrated a considerable countercampaign in support of the *Bullion Report,* largely through the *Quarterly Review.*[46] When Greenfield had reviewed Ricardo in February 1810 he expressed the feeling that "[t]he views of Mr. Ricardo on this subject are . . . consonant with our own."[47] But it was only in late

1810, with the emergence of the Canningite campaign, that the issue really came to the fore in the *Quarterly Review.*

In prefacing his pamphlet Huskisson stated that it was with deep regret that he was witnessing

> an attempt made to create political divisions on the subject [the Bullion Report]; and to array particular parties against *principles* which, surely, are not to be classed among the articles of any political creed, or to be considered as connected with the separate interests of any party.[48]

Interestingly Huskisson also noted another level of agitation quite explicitly. He wrote, "The speech of Mr. Randle Jackson, though it imputes *party spirit* to others, is obviously dictated by nothing more than *corporation spirit:* a distinction which, fortunately, is too plain to be misunderstood."[49] Clearly Ricardo could see it, and our cultural theory-inspired analysis of the correlation of cultural bias and social relations thus far alerts us to it. However, it appears to have dropped out of the sight of many historians of economic thought who have pursued conventional historiographic approaches.

Huskisson began his substantive argument by taking a characteristically, indeed definitionally bullionist stance. Declaring the equivocal sense of the term *money* to be a point of confusion in the current debate, Huskisson asserted that money was in fact not merely conventional but must also possess intrinsic value.[50] It must be an equivalent, and not merely a representative of value. The crux of Huskisson's argument lay in outlining the various current laws relating to money, including the Restriction Act, with the aim of showing that gold to a specific quantity and quality was the standard of value of currency, and that it was no part of the intention of the legislature in enacting restriction to vary that standard, and yet that it had in fact been varied by the actions of the Bank, and that, therefore, banknotes were depreciated.

Huskisson held out the nonsensical state of the laws in 1810, which allowed a light guinea, which could be melted legally, to be worth more in bullion than was a full-weight guinea containing by definition more gold but which could not be legally melted,[51] to be the manifestation of the contradiction introduced by the actions of the directors of the Bank of England under the cover of restriction. It was to prove a central element in the argument of 1819, that the intentions of the legislature had been perverted, but in 1810 it was little noted.

In essence the *Quarterly Review* writers argued that restriction had been a necessary emergency measure in 1797, but that it should not stay in force when no longer necessary. In the *Quarterly Review* of that campaign period there was a shift toward suggesting that whatever had been the cause of the gold price and exchange problems, the solution lay in the reduction of the amount of Bank paper in circulation.[52] That is to say that there was a trend toward accenting the recommendations and conclusions of the *Bullion Report* over and above the details of causal analysis—the principles.

There were personal issues involved in this refocusing away from the original cause. As a prominent member of the Treasury at the time of restriction in 1797, Huskisson can hardly be expected to have done other than to play down the issue of culpability. Huskisson accounted for his change of mind between 1797, when he was in office, and 1810, when he was so recently put out of office, by saying that he had only recently realized that the Bank of England directors followed no sound principle in their accommodation.[53] He had only realized that the Real Bills Doctrine ruled during the Bullion Committee's inquiry.

The socio-cultural aspects are also of interest. William Huskisson was not of an aristocratic family. Indeed the family acquired land only gradually, and Huskisson's chief access to independent means was through his marriage. After schooling in England and France Huskisson became private secretary to Lord Gower in Paris during the late 1780s and early 1790s. Huskisson was present when the Bastille was taken,[54] and was an early supporter of the revolution. He soon became disenchanted, indeed horrified, by the violent excesses that followed. Huskisson was member of parliament for Chichester from 1812 until 1823, and then represented Liverpool until his tragic death in 1830.[55] The family home base was Staffordshire, the increasingly industrial black-country, and Huskisson's "constituency" was the industrializing northwest of England. Intellectually he is said to have been a very talented follower of Adam Smith.[56]

William Huskisson was not an old-school Tory with landed aristocratic lineage. He was more of a functionary of government, serving as secretary to the Treasury and in the Board of Trade. His personal situation was, after his inheritance and marriage, one of independence rather than of wealth. His intellectual lineage was from Adam Smith, and his great political and financial hero was William Pitt. Hence, one can see many of the socio-cultural features that would have led Huskisson to play a pivotal role. Like Ricardo, he was in the important central grid-group-power region. He was not, in the 1808–21 period, in a highly powerful position, nor was he ever subject to the great burden of the obligations of aristocratic life. As a functionary he was subject to bureaucratic experience, and can, therefore, be placed toward the lower end of the high grid region. With independence Huskisson was close to the central power and group regions. (See Figure 5.2.) So one

Figure 5.2. The Position of William Huskisson (1810–1812)

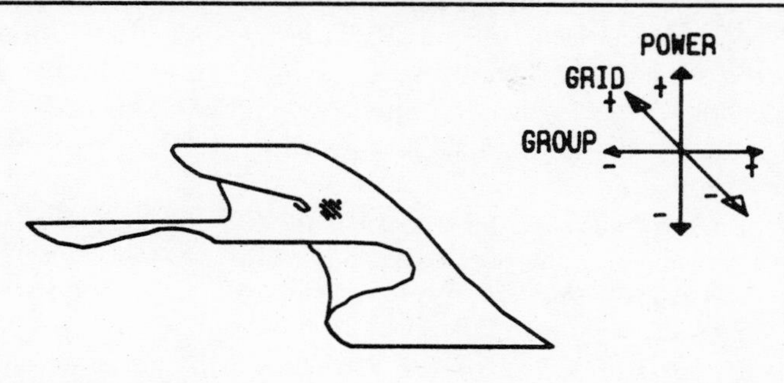

⁂ Indicates position.

Source: M. Thompson, 1982, 'A Three-Dimensional Model,' In Mary Douglas, Ed., *Essays in The Sociology of Perception* (London: Routledge & Kegan Paul), p.50.

can place Huskisson toward the center, but just inside the hierarchist region. Huskisson's monetary thought can be seen to reflect this position, as does his role in monetary debate. Of the hierarchists he was closer to the bullionists than others. Like Ricardo he was, in the revealing dual sense emergent from socio-cultural analysis, central.

Within the framework of this refocusing the *Quarterly Review* warned of the deflationary effects and serious consequences of too rapid an adjustment. The reviewers championed resumption as the institution of an automatic system of adjustment over any method or rule of principle, be that principle Real Bills or any other. The crux for them lay in protection against inflation, not in the current fashion in principles.[57] As Tories, albeit oppositionists and, relatively speaking, economic liberals, the Canningites cannot be expected to have wholeheartedly adopted a style of thought based on principles. For such men, for men of such social experience, the attraction of bullionism lay not in principles, but in its offering an automatic system which avoided conflicts between private interests and public duty. Nor, indeed, can they have been expected to have accented *cause* above monetary *order.*

The Whig bullionists were, of course, delighted that Canning and Huskisson, two prominent Tories, should side with the Bullion Committee's conclusions, since it destroyed the arguments of Jackson and Sinclair, that the whole thing was Whig plot. This is no more than saying that the quarterly reviewers hit their target. The Canningites of the *Quarterly Review* held out the carrot of compromise insofar as they moved away from both the Tory-versus-Whig alignment *and* sought to move away from attributing blame. They concentrated on the cure: the recommendations of the *Bullion Report.* Tory opponents of the report were not, however, to be easily tempted. Prime Minister and Chancellor of the Exchequer Spencer Perceval summed up the position in a letter to Huskisson on 30th October 1810, when he wrote:

> Neither the reading of the report nor the Circumstances which have since occurred have to any degree diminished my *apprehension of the ruinous consequences of what the report recommends.*[58]

Perceval, as a member of the old Tory elite, was worried about the consequences, about keeping an order, not about the *principles* involved.

During 1810 consciousness of the issues of currency and of the *Bullion Report* had grown to a crescendo; pamphlets and articles poured from the presses. Campaigners on both sides of the controversy as well as concerned mavericks rushed to present their cases before the public, and before the members of parliament on the whom the responsibility for decisive action on the basis of their analyses of the situation rested. In February 1811 the *Edinburgh Review* featured an extended review by Thomas Malthus of what he implied were some of the most important contributions to the debate.[59]

Malthus began by congratulating Mushet and Ricardo on the timeliness of their publications, which he said had "given a beginning to the interesting discussion which is still going on."[60] Of Ricardo's *High Price of Bullion* Malthus said that it pointed out two great truths: first, that "every kind of circulating medium, as well as every other kind of commodity, is necessarily depreciated by excess, and raised in value by deficiency, compared with the demand, without reference to confidence or intrinsic use," and second, that "excess and deficiency of currency are only *relative* terms, that the circulation of a country can never be superabundant, except in relation to other countries."[61] Malthus concluded that "both he [Ricardo] and Mr. Mushet appear to us to have completely succeeded in proving the actual depreciation of our currency, and in tracing it to its true cause."[62] Having said this, however, Malthus made it clear that he thought that the great fault of Ricardo's pamphlet was his failure to recognize that the state of the exchanges could be the effect of real trade factors as well as of purely monetary ones.

Malthus's treatment of Bosanquet was rather summary. He declared it impossible to take seriously an analyst whom he found "on second thoughts, *giving up completely the question of depreciation,* in reference to our legal tender,—*acknowledging fairly* that the gold contained in a guinea is now of more value."[63] In the light of Bosanquet's second thoughts Malthus saw no reason to argue, since he had in fact conceded defeat. He tellingly referred to Bosanquet's effort as "a most unlucky specimen of the reasoning of practical men."[64] Malthus thus not only criticized the argument, but also recognized the style of thought, and placed it in context.

Taking the occasion of the review, and having the attention of such a wide audience as the *Edinburgh Review* enjoyed, Malthus embarked on an extended series of remarks of his own. In our context, however, it is perhaps his analysis of the state of the controversy in discursive and socio-political terms that is more interesting than his monetary thinking. Malthus made two telling observations, one concerning the mode of bullionist argumentation and the other concerning the political position of the Bank of England. He wrote:

> One of the principal faults which we have remarked in almost all the writers that are unfavorable to the Bank restriction, is, that they have not made sufficient concessions to the mercantile classes in some points where they appear to have the truth on their side. . . . A practical merchant must, to be sure, be extremely surprised at such a denial, and feel more than ever confirmed in his preference of practice to theory.[65]

At the same time Malthus indicated his awareness that merchants had a vested interest in the greater facility of loans, and in the rising prices that restriction had brought.[66] On this second point Malthus wrote:

> We were, at first, inclined to approve of the recommendation of the [bullion] committee, to leave to the knowledge and discretion of the Bank Directors the mode of preparing themselves to resume their payments in cash at the time proposed. But it has been suggested, and the language and conduct of their friends have not sufficiently repelled the suspicion, that, under cover of this liberty, they might purposely keep the same, or a greater quantity of notes in circulation, with a view of compelling the legislature to continue the Restriction Act, as there would, of course, be a great unwillingness in all quarters to enforce a law which at the time could not be obeyed, and the attempt to obey which, in such a state of things, would produce very serious inconveniences to the public, as well as to the Bank. . . . It is, indeed, an monstrous deformity in the state, that an incorporated body of individuals should have the power of holding out a threat to the legislature, that if it does not persevere in sanctioning the nonfulfilment of their engagements, they would find a means of embarassing and punishing the government and the public.[67]

In the light of this analysis of the situation in the City Malthus recommended that the legislature must *enforce a plan* on the Bank if it really wanted resumption to happen.

In response to the *Edinburgh Review* article Ricardo wrote an appendix to the fourth edition of *The High Price of Bullion,* which appeared in April 1811.[68] This short appendix contained a number of important points, although it has commonly been seen as no more than a first approximation of a plan for resumption. One point that Ricardo made was to draw a distinction between an increase in banknotes and a redundancy of currency. In an attack on the review and on the evidence of Mr. Pearse, the deputy governor of the Bank of England, given before the Bullion Committee—in which it was stated that "[t]he varying prices of the Hamburgh exchange compared with the amount of bank-notes at different periods, seem to prove that the amount of bank-notes in circulation has not had an influence on the exchange"[69]—Ricardo pointed to their error. They have, he said,

> wholly mistaken the principle advanced by those who are desirous of the repeal of the restriction bill. They do not contend, as they have been understood to do, that the increase in *banknotes* will permanently lower the exchange, but that such an effect will proceed from a redundant currency.[70]

He went on to cogently argue that there was no necessary connection between an increase in banknotes and a permanently increased and redundant currency. Interestingly, and most importantly, Ricardo achieved this by means of an analysis of the institutional distinctions between notes of 5 pounds and over, notes of less than 5 pounds, and coins—*not by the route of a velocity of circulation argument.* Ricardo provided a means by which to reconcile the facts, as presented by Pearse, with his principles, a means by which to render the bullionist argument *independent* of the analyses of Thornton and Horner, who were, at least up until the time of the publication of the *Bullion Report,* considered to be defenders of the Bank Restriction Act. It was important, some say disastrous,[71] for the later development of monetary theory that Ricardo was able to maintain his simple quantity theory.

In this way the eventual triumph of bullionism came to involve a retrograde movement in the development of monetary theory. This demonstrates the victory of a weaker theory over a stronger one. It also gives an instance of the context-related development of analytical details, which are *not* logically necessitated by the conclusions or theory structure. The mainstream approaches, associated with the philosophers of science Imre Lakatos and Karl Popper, involve either a rational reconstruction of the history of science wherein the stronger theory and/or the theory with greater empirical content wins out, or a logical structure wherein that which explains (explanans) logically entails that which is explained (explanandum). The triumph of Ricardo's simplistic quantity theory based on principles rather than empirical complexity, and on the quantity theory without the velocity of circulation argument, fits neither of the mainstream models of theory choice.[72]

Following the publication of the *Bullion Report* William Cobbett embarked on an extended history of the controversy that had led up to its production in a series of 29 letters from Newgate Gaol, entitled "Paper Against Gold; Being an Examination of the Report of the Bullion Committee" in the *Political Register* (Volumes 18 through 20). In prefacing this series of letters Cobbett wrote of the *Bullion Report:*

> This Report has given me more delight than anything I ever set my eyes on, my wife and children only excepted. I am in prison, to be sure, but my principles are *at large.*[73]

It may, however, have been more accurate for him to have said that his recommendations were at large, rather than his principles. Cobbett joined enthusiastically in the cry for resumption,[74] but he referred to the level of commodity prices as the mark of depreciation. In this he was not in agreement with what was characteristically, indeed definitionally bullionism.

Despite extensive wandering around the issue a core line of analysis did emerge from Cobbett's correspondence. Central was his analysis of the national debt. Cobbett suggested that "banknotes [had] increased with the [National] Debt,"[75] and that the only way to reduce the quantity of paper was to reduce that debt. The main target of blame was thus shifted by Cobbett to the veiled dishonesty of Pitt's sinking fund. Cobbett argued that the sinking fund, begun in 1786, was meant to be a means of reducing the debt, whereas, in fact, it had been used to merely repatriate it. Tax was levied in order to buy back part of the debt, but that debt was not in fact cancelled. The tax was used to pay interest on it, and as further input to the sinking fund. Thus the government had progressively bought back part of its debt, but at no stage had it actually cancelled or redeemed any part of it.[76] It was his analysis of the national debt, the system of finance, which allowed Cobbett to argue the case for gold against paper. He suggested that real money represented the worth of things, while promissory notes represented debt.[77] Against such champions of the paper money system as Jackson and Sinclair he asserted that banknotes were produced by increased debt, *not* by increased wealth, as they argued.[78]

Cobbett related the whole to his pet issue of parliamentary reform, and thus placed *both* the Tory and Whig arguments in that wider political context. The cost to the nation of sinecures and placemen could only be guessed at, but Cobbett argued that all great sums are made up of small sums. Cobbett proved that, contrary to the claims of the defenders of the paper money system, it was *not* a necessity of war.[79] In the end, and despite his own former vocal opposition, Cobbett became a great champion of Thomas Paine's arguments on the paper system of finance. "Paine's twenty-five pages," Cobbett wrote, "convey more useful knowledge on this subject, and discover infinitely greater depth of thought and general powers of mind, than are to be found in all the pamphlets of the *three score and two* financiers, who, . . . [have] favored the world with their opinions."[80]

Cobbett's analysis is interesting because it shows another, wider context of interestedness, which supplements the party political and corporate/City contexts, and another way in which monetary debate could be turned to serve a cause. As an outsider to the City factionalism Cobbett was in a position to think of, to see, an alternative way of thinking about the financial features of the period. Cobbett was an old-style reformer who sought not a new order, but a return to an idyllic past order. Cobbett's vision was of an agrarian utopia, an egalitarian simple life that had never really existed. He campaigned against the Tory elite's abuse of position and for the equality of access and of rights. Like many radicals of the time Cobbett was an egalitarian. He was left behind by the times. In terms of socio-cultural analysis, he was in neither of the key regions. Consequently his analysis did not present a viable alternative.

During late 1810 the supporters of the *Bullion Report* generally had the best of the exchange, but as the day of the debate of the *Bullion Report* in parliament grew nearer there was a renewed surge of pamphleteering in which the tables were somewhat turned. In an 1810 issue of the *Monthly Review* the reviewers wrote:

> We have [reviewed] five vigorous assailants, and only one solitary defender, of the bank-note system; and even this proportion is more favorable to it, we believe, than would be found on taking a more comprehensive estimate of public feeling.[81]

Yet they wrote in the issue of March 1811: "Since the late increase of these performances, it deserves remark that the balance, in point of number, has changed sides, and is now in favor of the Bank."[82] What is also clear is that the reviewers saw the sides as Bank versus anti-Bank factions, not in terms of principles. As the pro-restriction campaign blossomed and attained the ascendancy in terms of quantity, however, the two elemental groups in the campaign began to show the first signs of a split.

In his *Considerations on Commerce, Bullion and Coin, Circulation and Exchanges* George Chalmers, a Board of Trade official, presented his ideas on the points of exchanges, the price of bullion, banknotes, and Bank affairs. On exchanges, Chalmers asserted that trade disruption was the major problem, and that the true restorative was the increase of exports. One might have expected the Board of Trade to promote trade. Chalmers, somewhat inconsistently, went on to attribute the high price of bullion to its scarcity. On Banks, Chalmers's key thrust was an hostility toward country banks. He went so far as to describe notes issued by country banks as "an excrescence on our commercial dealings."[83] This was in complete contrast, of course, to the "sound and salutary" issues of the Bank of England—echoes once more of the evil outside/good inside rhetoric of persons with high group commitment in the context of the factions of the City. Chalmers was clearly pro-Bank.

In *Observations on the Fallacy of the Supposed Depreciation of Paper-Currency of the Kingdom* (1811) F. P. Eliot launched an attack on Huskisson regarding the issue of depreciation. This mode of attacking Huskisson's pamphlet became quite widespread as the pro-restriction campaign developed. Eliot argued that "the pound sterling, in money of account, is our only accurate and invariable measure of wealth."[84] Gold was not the measure, but was merely another commodity. By means of adopting this abstract standard Eliot was able to breeze through the question of depreciation. He suggested:

> There is no comparative depreciation between our paper and metallic currencies; that there is no unneccesary augmentation of the circulating medium; that the annual supply of gold from the mines is not equal to the increasing demands of the world for that metal; and that the increase of paper-currency has barely supplied the place of the augmentation wanting in the metallic portion of the circulation.[85]

The notion that paper was at par so long as it exchanged for coin was a delusion into which many pro-restriction writers entered, despite the best efforts of almost all the serious analysts to show that *all currency*, coin and paper, depreciated together when it was inconvertible. It was yet another manifestation of the failure to distinguish a situation of convertibility from one of inconvertibility.

Robert Wilson's *Observations on the Depreciation of Money, and the State of Our Currency* (Edinburgh 1811) took a very strong line on trade as *the* cause of the exchange problems. He went so far as to assert that "a reduction in our notes *never* can be the means of restoring our exchange."[86] Of depreciation Wilson suggested that the Corn Laws were a principle cause, and that taxation was a subsidiary one.[87]

Such works prefigured the subsequent rift between Bank and government, specifying their very particular interests. To cite trade as a/the cause was to blame Napoleon—the common enemy without; to cite taxation was to blame the government, while to cite overissue was to blame the Bank. In 1810–11 it was possible to blame Napoleon, but in the later debate, with Napoleon no longer around, analysis was to devolve into the Bank and government blaming each other.

A speech by Stephen Cattley at the court of the Bank of England on 21st March 1811 brought the issues and the pro-Bank line of reasoning home. Cattley asserted that the way to cure the "present evils"[88] was to restore the exchanges. Note once more the rhetorical use of the notion of evil. He suggested that this might be achieved through trade means, specifically, cutting short imports. Having thus dispensed with the theoretical issues Cattley came to the heart of the matter by sharing his view of what the Bank would do in the event of a compulsory order to resume cash payments, namely, "demand from the government the six millions at present advanced on loan."[89] In the context of the parliamentary debate, and in the minds of the ministers, this was a key and all-too-material a threat. Like the emergent differences of analytical detail amongst the pro-restriction pamphleteers this too was indicative of the emergent rift between Bank and government, of a factional realignment.

After considerable delay Francis Horner moved parliament to discuss the *Bullion Report,* and the House of Commons resolved into committee for that purpose on 6th May, 1811.[90] While raising sixteen resolutions concerning the report, Horner registered his wish to separate the last resolution, that calling for resumption, from the rest, that is, to separate the questions of principles and analysis from ones of policy recommendations.[91] He hinted at a desire to get agreement over principles.[92] Horner argued, as Huskisson had done in his recent pamphlet, that it had not been intended that restriction should allow a deviation of the value of promissory notes from their legal standard, but that this had in fact happened due to an overissue of those notes.[93]

Henry Thornton also spoke of his desire that the House should agree on the principles at issue.[94] Thornton wanted to get the directors of the Bank to see and admit that the quantity of their paper had an influence on the price of bullion *and* the state of the exchanges.[95] He wanted to get parliament to officially agree that it was an undeniable principle that the quantity of bank paper regulated its value (other things being equal).[96] Like Horner, he sought to separate principles from recommendations, and to get agreement on the former first.

On the following day, 7th May 1811, Nicholas Vansittart put forward a series of sixteen counterresolutions. They merely stated the views of the directors of the Bank of England and the ministry in a positive way. Vansittart argued about the meaning of the term *depreciation.* He suggested that in fact depreciation meant that notes were not the same as legal money (coins) in transactions, and that notes buy less. He alluded to the fact that paper banknotes were still equivalent to legal coin for internal purposes,[97] implying thereby that paper and coin were equal. Indeed Vansittart's third counterresolution stated that the Bank of England's notes "are at this time, held in public estimation to be equivalent to the legal coin of the realm."[98]

Lord Castlereagh, Canning's great enemy, rose to suggest that the House should be debating policy and not the principles of political economy, and that it seemed to him that Horner, Thornton and Huskisson were not now debating policy.[99] "The House is," he said, being

> called on to decry the system of our currency, to stigmatise the Bank for erroneous, if not abusive administration of its functions, without having any distinct measure of correction suggested for their adoption.[100]

Once again it can be seen that the debate was between opposing styles of thought, which sought to pursue principles, on the one hand, and practical matters of fact on the other. Even in parliament it was a conflict of styles of thought, the styles associated with the individualist economic parvenus and the hierarchist Tory/merchant elite.

The next day Henry Parnell replied to Vansittart's resolution concerning the definition of, and thereby the fact of, depreciation. Parnell observed that Vansittart had agreed with the 1804 Irish Currency Committee that a depreciation of Irish currency had occurred because, on his own admission, guineas had exchanged for a premium in Ireland at that time. If this was the deciding test of depreciation, then the current state of the English pound was, suggested Parnell, a simple matter of fact, and not one of principle.[101] Parnell was able to refer to three cases of double pricing now being tried in the courts, and to the notoriety of widespread double pricing in the country. Moreover, with reference to the latest newspapers, Parnell reported that there was clearly a premium of 15½ per cent on guineas over Bank of England paper currently being given in Ireland.[102]

Notwithstanding Parnell's cogent argument, debate in the House of Commons led to the defeat of all of Horner's resolutions, and of the recommendations of the Bullion Committee. Resolution sixteen, seeking resumption in two years, failed by a vote of 45 to 180 on 11th May 1811.[103] Both the analysis *and* recommendations, theory and policy, of the Bullion Committee were rejected.

In subsequent debate Spencer Perceval argued that the relative prices of bullion and coin—the market and the mint price of gold—were more complex than the supporters of the report allowed. This was so because of the effects of the ban on the export of coin and melt of full weight coin, which the bullionists were always keen to set at nought. The fact was, declared Perceval, "that coin was not the same value abroad as bullion, because it could not be exported; and bullion was not the same value at home as coin, because it was not legal tender."[104] It was, of course, an argument in the style of empirical complexity, as opposed to the simplifying principles of bullionism. Alexander Baring got back to the crux of the matter, declaring that the real point was that a return to gold was *not possible.*[105] After consideration of the proposed amendments the whole of Vansittart's resolutions were agreed to on 15th May 1811. Thus the cause of resumption *and* the essential principles of analysis were all lost.[106]

Within parliament, as outside, the argument was conducted between holders of distinct styles of thought. Thornton and Horner sought an agreement as to principles, Huskisson sought agreement as to facts, and the old guard, such as Castlereagh, displayed their complete distaste for principles of any sort. The man on the spot, Prime Minister and Chancellor of the Exchequer Spencer Perceval, pointed to the empirical complexity of the case as opposed to the drive of bullionists, those pushing for resumption, toward simplifying principles. Bullionism, as principles and policy, was defeated because it was not, at that stage, a part of the dominant style of thought, the style of thought of the dominant group(s) and/or class(es). It was not the style of thought that the cultural bias of the social relations of the then politically dominant group could accept. Their way of life was supported by an alternative cultural bias wherein a complexity of gradations implied the need for a regulated hierarchy of obligations.

Following the decision in parliament to continue restriction, debate outside the House continued over both policy and principles. In what was, ostensibly, a review of five pro-restriction pamphlets in the *Edinburgh Review* of August 1811,[107] Malthus pursued the argument for resumption. Like Thornton, Malthus was not in any literal sense a bullionist, but he began to argue that whether the cause of the depreciation of currency was to be found wholly within trade factors, wholly within monetary factors, or in some combination was a subsidiary point. What mattered was that the fact of depreciation demanded that the value of the currency be restored.[108] This was a stance which began to gather increasing support, especially when the controversy resumed after the parliamentary defeat of bullionism in 1811.

In his review Malthus made much of Vansittart's resolution of 13th May. He pointed to the case of Lord King's rents as a clear demonstration that the estimation of equivalence did not, in fact, adhere—it was so only under legal compulsion.[109] Given that in the consequent Lord Stanhope's Bill there was official recognition of the fact of depreciation, it was nonsensical for the pro-restriction writers, the Bank directors and the ministers, to deny it.[110] And yet they did still deny it. In concluding his review Malthus wrote that nothing

> more directly contradicts the spirit of British legislation in the best times, than that which thus gives up 20 millions worth of revenue belonging to British subjects, to be regulated in its value according to the will and pleasure of 24 individual *merchants,* whose interests are in reality different from those of the owners of such revenue.[111]

In this and similar arguments the anti-Bank groups began to drive home the wedge which was splitting Bank and legislature. In later debate this trend proved to be most influential. Malthus also drew attention to what was to prove a more practical influence. As a means of escape from the predicament, he recommended that the plan outlined in the appendix to the fourth edition of Ricardo's *High Price of Bullion* be considered.[112]

After the defeat of the *Bullion Report* in parliament, debate and controversy over monetary matters soon died down. By the end of 1811 the flood of pamphleteering and campaigning had slowed to a trickle. It was not until after the Napoleonic wars that monetary controversy arose again.

AN ANALYTICAL SUMMARY

The analyses of Thornton and Horner were amongst the most theoretically nice and, perhaps, genuinely disinterested of that period. Nevertheless, the conclusions of the *Bullion Report* were loaded explicitly toward the "rights of commercial property"[113] and the restoration of the "natural system"[114] so dear to the less radical among the Whig reformers. Opposition to the committee's findings at the theoretical level consisted almost exclusively in the bankers' Real Bills Doctrine. This was buttressed by the Tory rhetoric of defense of King and Country, its practical political support. Given this rather weak opposition, one must ask why the Bullion Committee's principles *and* recommendations were not accepted.

It was widely suggested that the failure of the *Bullion Report* to secure parliamentary support was due in large part to the committee's apparent disregard for the evidence of witnesses, who were said to have concluded in a manner contrary to the committee. When one considers all the evidence, however, this perception seems less than entirely accurate—the evidence of Sir Francis Baring, and of the anonymous continental merchant Mr. ****, being cases in point. Moreover, it was more the speculative opinions of the witnesses than their reporting of the facts that the committee may be said to have ignored. The authors of the report appear to have used selective presentation of the evidence in the body of the report as a rhetorical device, presenting only that which they wished to refute. To those who read only the report this gave the impression of disregard for evidence, but the impression is somewhat dispelled if the collected and appended evidence is also read. It may have been fatally optimistic of the committee to assume that all members of parliament would read the complete documentation thoroughly.[115]

It is interesting to note that in his evidence the anonymous continental merchant, Mr. ****, stated it to be his opinion that the wartime disruptions to trade had *caused* the exchange depreciation, but that the failure of an automatic recovery was attributable to inconvertibility.[116] This point was little discussed at the time, but it does serve to show another way in which the causal analysis may have been completely dissociated from the conclusions. Mr. ****'s analysis embraced the anti-bullionist version of the operative cause(s) and the bullionist analysis of the cure. This throws into highlight the fact that the concerted effort by many bullionist writers to blame the directors of the Bank of England for overissue and depreciation was not a necessary part of winning the case for resumption, and that, conversely, it was not necessary to excuse the Bank to maintain restriction. Why then were analyses and remedy prescriptions associated as they were?

The clue, of course, lies in the fact that it was only the continental merchant, the outsider—outside the local group factionalist social context—who did, or could suggest an alternative association. Analytical structure went hand in hand with socio-cultural environment, and the monetary debate, the emergence and development of monetary theory and policy, was a confrontation of styles and modes of thought because, and insofar as, it was a confrontation of groups and/or classes with contrasting lived-experiences, commitments, and cultural biases.

The identification of this disjuncture has methodological implications. Analysts of the positivist ilk presuppose that the explanans logically entail the explanandum, that the conclusions follow from the analysis as a matter of logical necessity. They tend, therefore, to concentrate solely on the analysis. It can be seen that this leads to a partial and inaccurate history. Often there is no single mode of associating analysis and conclusions, but rather a tell-tale choice between different possible modes of association. This leads them to overlook the extent to which the content of knowledge, of analysis, is socially determined.

In highlighting the disjuncture between analyses and conclusions, then, we seek to refute the notion of a logically necessary relation between explanans and explanandum. Having done so, the question of the linking of the conclusions and analyses in some other way arises. It is suggested here that there are multiple (logically) coherent routes of analysis by which to arrive at any given conclusion or recommendation, and that the choice of route is socio-culturally determined. Thus it emerges that the relatively common linking of specific policy conclusions or recommendations with interests is a secondary concern. What is perhaps more important, though far less obvious to analysts adhering to epistemologies founded on logical structure, is the link between interests and analysis. What this suggests is the legitimacy of the strong sociology of knowledge; showing the impact of interests on the *analytical content* of knowledge.

The patterns of the styles of thought emerging in the bullion controversy have been traced, and they have been associated with both the real social environments of the actors, and with the development of opposing theoretical perspectives. By noting that alternative analyses had their source in *outsiders'* social environments we have been able to tie the modes and forms of analyses to their social environment. During the course of the debate circa 1810–11, certain key evidence that might be said to lend support to our claim for the usefulness and viability of the socio-cultural schema has been noted. Following the publication of the *Bullion Report,* and a high degree of early acceptance thereof, there emerged a campaign of opposition to its recommendations and analysis, which was quite clearly interested. It was a campaign undertaken deliberately, often coordinated by Bank of England and Tory ministerialist operatives. Anti-ministerialists, both Whig and Canningite Tory, campaigned in support of the *Bullion Report.* So too did the anti-Bank of England forces from the City. In general the campaigns exhibited the evidence of factional motivations, and took the forms expected.

The exchange between Bosanquet and Ricardo is particularly revealing. Bosanquet, the son of a former governor of the Bank of England, adopted a mode of argumentation based on facts. It displayed just those characteristics of socio-cultural allegiance noted before in the works of Sir Francis Baring, namely, empirical complexity, goal of order, etc. Ricardo's reasoned analysis was, on the other hand, a most exemplary abstract argument, in which everything was reduced to a few simple relations or principles. That it was Ricardo who took it upon himself to respond to both Bosanquet and Jackson, the Bank's hired prize fighter, indicates the accuracy of our expectations. Ricardo was both furthest along the road to the emergent way of life and its cultural bias, and, of course, personally involved. He was, as has been hinted, the champion of the *Bullion Report* because he was the furthest along the road to socio-cultural commitment to its mode as well as its content.

Interestingly, however, it is not the heated debate circa 1810–11, but rather the silence from 1812 to 1816, that speaks loudest on the topic of City factionalism and its relation to the development of monetary theory and policy. The deaths of Sir Francis Baring and of the Goldsmid brothers left the Stock Exchange community in the ascendant. On 20th May 1811, just five days after the defeat of the *Bullion Report* in parliament, the Stock Exchange group of Barnes, Steers, and Ricardo, the Gentlemen of the Exchange, gained a share of the government loan of that year. They were the successful loan contractors for the first time since 1807. The remaining elements of the old-guard faction in the City, Battye, Baring etc. were the unsuccessful contractors.

Having established a system of loan contracting through the Stock Exchange, and their own position in the City, the members of the Exchange seemed to forget their differences with the merchant and banker groups, and, interestingly, their cry for free market competition in loan contracting. For the three government loans from 1812 to 1814 they colluded with their former enemies and submitted a single bid.[117] In apparent free competition for the loan of 1815, all four contracting syndicates submitted an identical bid, which was also, singularly enough, the exact minimum which the Treasury had resolved to accept.[118] What is remarkable in the context of the bullion controversy is that the 1812 to 1815 period, in which the formerly warring City groups apparently resolved their differences, *exactly coincides with a marked lull in the debate.*[119] That is to say that the sudden cessation of factional hostilities, implying down group movement, coincided with an analytically predictable lull in monetary debate.

This calm befell the debate despite the fact that the maximum gold premium of almost 50 per cent was reached in *July 1813.* (See Table 4.3 above.) This suggests that the controversy was not linked directly, or simply, to economic events. Debate arose for reasons other than the mere presence of a problem. What is clear is that (positivist) attempts to link debate with events fail in this case, just as they failed in the case of the Irish bullion controversy. (See Chapter 3 above.) Far from relating to economic events or crises, the theoretical and policy debates

Figure 5.3. The Position of the Economic Parvenus (1812-1814)

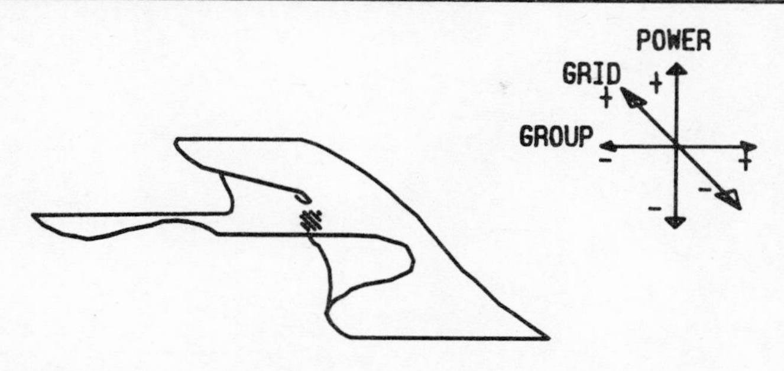

⋇ Indicates Position.

Source: M. Thompson, 1982, 'A Three-Dimensional Model,' In Mary Douglas, Ed., *Essays in The Sociology of Perception* (London: Routledge & Kegan Paul), p.50.

arose out of, and were an aspect of, the underlying social relations of power.[120]

The socio-cultural schema alerted us to the centrality of factionalism, *and* to the likelihood that, after a period of clarification of positions, there might be some tendency toward reconciliation in order to avoid a loss of power. The exact context of this theoretically hypothesized change is now revealed. Whereas the Gentlemen of the Exchange had sought status and power through differentiation and specialization within the London money market(s), and had thereby engaged the group dynamics causally active in the emergence and development of a new form of monetary theory, having gained the ascendency, they now sought to consolidate and extend their influence through reconciliation. This for two reasons: first, factionalism and group orientation is the means of the underdog, not that of the ascendant; and second, to extend influence it is necessary to move to a wider arena than that of the group. This must involve some loss of the rule-setting position, in the form, in this case, of the high degree of internal self-regulation that the economic parvenus, especially the Gentlemen of the Exchange, had enjoyed. In orienting and moving into the wider social arena their lived experience became somewhat less group centerd, and they experienced a loss of regulative control and of power. In short, there was an up grid, down group and down power movement. (See Figure 5.3.)

This development has some very interesting implications for the course of the later debate. Those amongst the Tory leadership began

Figure 5.4. The Relative Positions Approaching (1812-1814)

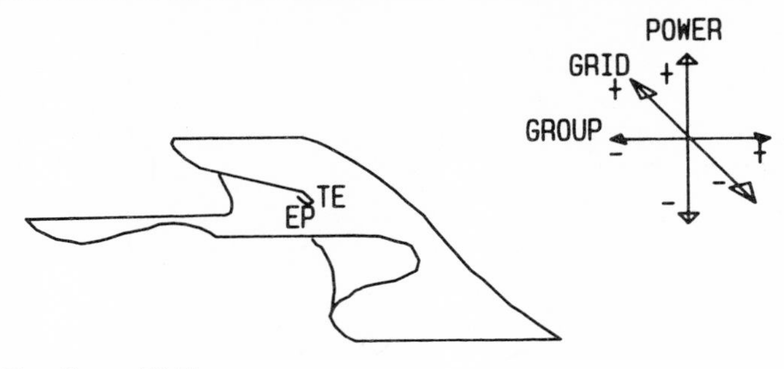

Where TE = The Tory Elite
and EP = The Economic Parvenus

Source: M. Thompson, 1982, 'A Three-Dimensional Model,' In Mary Douglas, Ed., *Essays in The Sociology of Perception* (London: Routledge & Kegan Paul), p.50. Reprinted with permission.

from a high grid and high power (hierarchist) position, but have by now experienced declines in power and, perhaps, grid prescription. Having succeeded in defeating the *Bullion Report* the Tory and Bank leaders may have also experienced some down group movement. As the campaign ended, the troops were demobilized and no longer unified from without. The economic parvenus began from a high power, low grid and low but ascending group (individualist) position, but are now on an up grid and down group trail. The implication is not only that the parvenus are shying away from the brink of the cusp region and moving instead on a continuous plain, but also that the two groups are moving toward rather than away from one another. (See Figure 5.4.) In view of these movements some evidence of a more gradual process of persuasion, and some theoretical and analytical reconciliation in the ensuing debate might be expected. We might expect social reconciliation or rapprochement to lead to analytical reconciliation or rapprochement. As the socio-cultural positions approach, the apparent socio-cultural and theoretical incompatibilities decline.

What does all this tell us about the motivation and intention of Ricardo and the "three score and two financiers" who favored the world with their opinions on monetary issues? Ricardo's polemical cry of disastrous consequences, and a situation "pregnant with present evil and future ruin"[121] is indicative of high group commitment. Clearly the timing of events does suggest that the factionalism amongst the economic leaders played a key role in the motivation and timing of controversy, much more so than did the mere presence of problems and sequence of (empirical) events. It is a most notable feature of the lull in the debate 1812–15 that the maximum depreciation and peak gold price were reached in 1813, but attracted no comment at that time. Both action and inaction attest to the correlation of factional and analytical conflict. More importantly, the impact of factionalism on the modes of analysis, on the associations of analyses and conclusions, and on the choice between theories has been revealed. But how does this analysis help historians overcome difficulties of interpretation? To answer this we might look briefly at the, so-called, Silberling thesis.

It was the central political thesis of Professor Silberling's 1924 paper *The Financial and Monetary Policy of Great Britain During the Napoleonic Wars; Part II, Ricardo and The Bullion Report* that, "behind the bullionist school is a bias to personal interests which distorts their analysis."[122] Silberling attacked Ricardo, arguing that his publications arose from a desire to further his fortune as a leading broker-jobber on the London Stock Exchange. According to Silberling, Ricardo's attacks on the Bank of England were intended to depress security prices, and to restore the profitability of the former continuation interest system.

Of the Silberling thesis it might be said that perhaps the head of the syndicate of the new-guard Gentlemen of the Exchange saw the opportunity to down the leading lights of the old guard, merchants and bankers—the Pillars of the City. People seek power as much as they seek riches. Indeed they often only seek riches as a means to power. That Ricardo translated his riches from the funds into land suggests that by 1813–14 his desire to consolidate was stronger than his desire to multiply riches. That he subsequently *bought* a seat in parliament is suggestive of an interest in power and influence, and of his using riches as a means to that end. Silberling and Sraffa, together with those who have weighed in on either side of the Silberling thesis argument, have founded their analyses of interested motives upon an assumed profit motive. For many historians with an economic leaning there is a tendency to equate interests with profit motive. This case, and it is hoped this study, demonstrates how limiting such a presupposition is, and how it handicaps such studies. Bertrand Russell wrote:

> Economics as a separate science is unrealistic, and misleading if taken as a guide in practice. It is one element—a very important element, it is true—in a wider study, the science of power.[123]

Ricardo can, we think, be cleared of the charge of the headlong pursuit of another pound. But can he be cleared of the pursuit of power, and the wish to help his friends in the Stock Exchange who had accepted him into their community as a young man, just as he was rejected by his family and was rejecting his Jewish heritage, and had made him their leader and principal hero? It must be remembered that the Stock Exchange community was a small and, because of the predominance of social and religious outcasts, a very tight-knit one. Anyone brought up within such a community would surely be deeply committed to it. It is suggested that Ricardo was so committed. What is clear, in any event, is that we can hardly justify attempts to treat the development of monetary theory as a purely intellectual process. The interrelationships between principles, practices and purposes are far too complex for that to be credible if the intention is to be descriptive.

Having looked at some of the factors involved in the failure and defeat of bullionism in 1811, we now turn our attention to its success in 1819.

NOTES

1. F. W. Fetter, 1959, "The Politics of the Bullion Report," *Economica,* Vol. XXVI, pp. 108–09.
2. B. Hilton, 1977, *Corn, Cash and Commerce; The Economic Policies of the Tory Governments 1815–1830* (Oxford: Oxford University Press), p. 46.
3. Ibid., p. 38, and Fetter, 1959, op. cit., p. 109.
4. Fetter, 1959, op. cit., p. 111.
5. J. Hill, 1810, quoted *Monthly Review,* Vol. 64, p. 280.
6. Earl of Rosse, 1811, quoted *Monthly Review,* Vol. 65. p. 321.
7. Ibid., p. 322.
8. *Hansard's,* 1811. Vol. XIX, p. 1008.
9. Sir John Sinclair (1754–1835). A lawyer and student of Edinburgh and Glasgow universities, he enclosed several thousand acres around Caithness in Scotland and was known for the introduction of the Cheviot breed of longwool sheep. He was the author of the authoritative *Statistical Account of Scotland* in the 1790s.
10. Fetter, 1959, op. cit., p. 112.
11. John Sinclair, 1810, quoted *Quarterly Review,* Vol. 4, No. 8, Art. xv, p. 520. Emphasis added.
12. *Quarterly Review,* 1810, Vol. 4, No. 8, Art. XV, pp. 518–536, and 1811, Vol. 5, No. 9, Art. VII, pp. 120–138. Both ostensibly by George Ellis.
13. *Quarterly Review,* 1811, Vol. 5, No. 9, Art. vii, p. 127.
14. *Monthly Review,* 1810, Vol. 63, p. 299.
15. Ibid., p. 300.
16. Ibid., p. 301.
17. *Hansard's,* Vol. XX, p. 168.
18. Idem.

19. Sinclair, 1810, quoted Cobbett, *Political Register,* 1811, Vol. 19, No. 42, p. 1312.

20. In chapter 6 it will be noted that Huskisson's argument that the rules had been perverted by restriction was a major factor in the conversion of traditional Tories in 1816. That is, that their conversion depended in large part upon an appeal to, and in the mode of their style of thought.

21. P. Sraffa and M. H. Dobb, 1951–55, Eds., *The Works and Correspondence of David Ricardo* (Cambridge: Cambridge University Press), Vol. III, p. 9., and F. W. Fetter, 1959, op. cit., p. 113.

22. Randle Jackson (1757–1837). Graduated from Oxford and was called to the bar in 1793. He acted as parliamentary counsel for the East India Company and the Corporation of London.

23. Sraffa and Dobb, 1951–55, Eds., op. cit., Vol. III, p. 146.

24. Ibid., p. 151.

25. Ibid., p. 147.

26. Idem.

27. P. R. Hoare, 1811, *An Examination of Sir John Sinclair's Observations.*

28. Hoare, 1811, quoted *Monthly Review,* 1811, Vol. 65, p. 319.

29. Charles Bosanquet (1769–1850). He was a member of a prominent Huguenot family of merchants, sometime chairman of the Exchequer Bills Office, Governor of the South Sea Company, and son of a former Governor of the Bank of England. Refer; Fetter, 1959, op. cit., p. 112.

30. Sraffa and Dobb, 1951–55, Eds., op. cit., Vol. III, p. 10.

31. *Quarterly Review,* 1811, Vol. 5, No. 9, Art. xi, p. 246. Emphasis added.

32. Charles Bosanquet, quoted Sraffa and Dobb, 1951–55, Eds., op. cit., Vol. III, p. 160.

33. Sraffa and Dobb, 1951–55, Eds., op. cit., Vol. III, p. 159. Emphasis added.

34. *Monthly Review,* 1811, Vol. 64, p. 173.

35. Ibid., p. 174.

36. Bosanquet, 1810, quoted *Monthly Review,* 1811, Vol. 64, p. 175.

37. *Monthly Review,* 1811, Vol. 64, p. 175.

38. Ibid., p. 177.

39. Ibid., p. 183.

40. Idem.

41. Ibid., p. 175.

42. *Quarterly Review,* 1811, Vol. 5, No. 9, Art. xi, p. 246.

43. William Huskisson, in J. R. McCulloch, 1966, Ed., *A Select Collection of Scarce and Valuable Tracts and other Publications on Paper Currency and Banking* (New York: A. M. Kelley), pp. 566–668.

44. Fetter, 1959, op. cit., p. 115, fn 2.

45. Idem.

46. Ibid., p. 116.

47. *Quarterly Review,* 1810, Vol. 3, No. 5, Art. xii, p. 156.

48. Huskisson, 1810, op. cit., p. 573. Emphasis added.

49. Ibid., p. 574.

50. Ibid., p. 579.
51. Ibid., p. 588–89.
52. *Quarterly Review,* 1811, Vol. 5, No. 9, Art. xi, p. 259.
53. Huskisson, 1810, op. cit., p. 574–77.
54. C. R. Fay, 1951, *Huskisson and His Age* (London: Longmans, Green and Company), p. 259.
55. The one thing that everyone knows about William Huskisson is that he was run down by Stephenson's *Rocket* at the opening of the Liverpool-Manchester railway in 1830.
56. Fay, 1951, op. cit., and G. S. L. Tucker, 1976, *William Huskisson Essays of Political Economy* (Canberra, Australia: Australian National University).
57. *Quarterly Review,* 1810, Vol. 4, No. 8, Art. x, pp. 414–453.
58. Fetter, 1959, op. cit., p. 116. Emphasis added.
59. *Edinburgh Review,* 1811, Vol. XVII, No. XXXIV, Art. V, pp. 339–72. Malthus reviewed; Ricardo (Jan 1810), Mushet (Jan 1810), Blake (May 1810), Huskisson (Oct 1810), and Ricardo (Dec 1810).
60. T. R. Malthus, 1811, *Edinburgh Review,* Vol. 17, No. 34, Art. V, p. 340.
61. Ibid., p. 341.
62. Ibid., p. 342.
63. Ibid., p. 357.
64. Idem.
65. Ibid., p. 361.
66. Ibid., p. 371.
67. Ibid., p. 370.
68. Sraffa and Dobb, 1951–55, Eds., op. cit., Vol. III, p. 11. It also appeared as a separate pamphlet.
69. Bullion Committee, 1810, "Report on the High Price of Bullion, with minutes and evidence," in S. J. Butlin, 1951, Ed., *Report on the High Price of Bullion, with minutes and evidence* (Sydney: The University Press), p. 85.
70. Sraffa and Dobb, 1951–55, Eds., op. cit., Vol. III, p. 114.
71. R. S. Sayers, 1953, "Ricardo's Views on Monetary Questions," in A. W. Coats, 1971, Ed., *The Classical Economists and Economic Policy* (London: Methuen), p. 37.
72. The case stands in refutation of logical-positivist and logical-empiricist inspired historiography.
73. William Cobbett, 1810, *Political Register,* Vol. 18, No. 5, p. 211.
74. Cobbett, 1810, *Political Register,* Vol. 18, No. 8, p. 260.
75. Cobbett, 1810, *Political Register,* Vol. 18, No. 10, p. 333.
76. Cobbett, 1810, *Political Register,* Vol. 18, No. 11, pp. 357–64.
77. Cobbett, 1810, *Political Register,* Vol. 18, No. 14, p. 456.
78. Cobbett, 1810, *Political Register,* Vol. 18, No. 15, p. 488. For a good exposition of this distinction see W. Santiago-Valiente, 1988, "Historical Background of the Classical Monetary Theory and the 'Real-Bills' Banking Tradition," *History of Political Economy,* Vol. 20, No. 1, pp. 43–63.
79. Cobbett, 1811, *Political Register,* Vol. 20, No. 3, pp. 71–73.
80. Cobbett, 1811, *Political Register,* Vol. 20, No. 1, p. 9.

81. *Monthly Review,* 1810, Vol. 63, p. 183.
82. *Monthly Review,* 1811, Vol. 64, p. 277.
83. G. Chalmers, 1811, quoted *Monthly Review,* 1811, Vol. 64, p. 284.
84. F. P. Eliot, 1811, quoted *Monthly Review,* Vol. 64, p. 288.
85. Idem.
86. Robert Wilson, 1811, quoted *Monthly Review,* Vol. 64, p. 291.
87. Ibid., p. 290.
88. S. Cattley, 1811, quoted *Monthly Review,* Vol. 65, p. 100.
89. Ibid., p. 100.
90. *Hansard's,* Vol. XIX, p. 798.
91. *Political Register,* 1811, Vol. 19, No. 37, p. 1149.
92. *Hansard's,* Vol. XIX, p. 799.
93. Ibid., pp. 831–32.
94. *Political Register,* 1811, Vol. 19, No. 39, p. 1213.
95. *Hansard's,* Vol. XIX, p. 895.
96. Ibid., pp. 896–902.
97. Ibid., p. 959.
98. Nicholas Vansittart, quoted Fetter, 1959, op. cit., p. 118.
99. *Hansard's,* Vol. XIX, p. 987.
100. Idem.
101. Ibid., p. 1036.
102. Ibid., p. 1037.
103. Ibid., p. 1069.
104. *Political Register,* 1811, Vol. 19, No. 41, p. 1299.
105. *Political Register,* 1811, Vol. 19, No. 43, p. 1336.
106. *Hansard's,* Vol. XX, p. 176.
107. T. R. Malthus, 1811, "Pamphlets on the Bullion Question," *Edinburgh Review,* Vol. 18, No. 36, Art. X, pp. 448–470.
108. Ibid., p. 451.
109. Cognizant of the issues at hand, and as a supporter of the bullionist cause, Lord King made an attempt to force the recognition of the fact of depreciation, and/or force the government to make Bank paper legal tender by attempting to insist that his tenants paid their rents in coin and not in depreciated banknotes. The case implied that double pricing, therefore depreciation, existed in fact. The Tory ministry were keen to prevent such activities, and they sought to enforce estimation of the value of paper currency through legislation. Lord Stanhope's Bill was the response to the ploy of Lord King, and it made it illegal to take or give gold coin or notes at anything but their face value. To the extent that it was necessary it undermined the arguments of the pro-restriction group(s). Refer; *Political Register,* Vol. 19, No. 52, and Vol. 20, No. 1.
110. T. R. Malthus, 1811, *Edinburgh Review,* Vol. 18, No. 36, Art. X, p. 449.
111. Ibid., p. 467. Emphasis added.
112. Ibid., p. 470.
113. Bullion Committee, 1810, in McCulloch, 1966, Ed., op. cit., p. 455 and p. 470.

114. F. W. Fetter, 1965a, *The Development of British Monetary Orthodoxy, 1797–1875* (Cambridge, Mass: Harvard University Press), pp. 53–54.

115. It is also interesting to notice how often the report has since been printed without the appended evidence—especially circa 1919–25.

116. Bullion Committee, 1810, in McCulloch, 1966, Ed., op. cit., p. 422.

117. E. V. Morgan and W. A. Thomas, 1962, *The Stock Exchange; Its History and Functions* (London: Elek Books), p. 50.

118. Sraffa and Dobb, 1951–55, Eds., op. cit., Vol. X, p. 82.

119. It is also notable that Ricardo's monetary *publications* suddenly ceased between April 1811 and February 1816, while his private correspondence continued to include monetary debate.

120. Parenthetically this 1812–15 period also reveals micro-sociological factors beyond those of the factions of the City already referred to. It is remarkable that Ricardo's life during this period of transition was also in transition. His change of personal circumstances brought an apparently complete and abrupt change of concerns and interests. In continuation of the correspondence on monetary issues Ricardo wrote to Malthus on 1 January 1814 enclosing a table of exchanges which he reported that he had lately been working on. Ricardo also mentioned his intention to visit Gatcomb Park during the coming week, his first visit, presumably in order to view the estate with the idea of purchase. While the sale of the property did not go through for some time, and Ricardo did not move there until July of that year, one assumes that the resolve to purchase must have been almost immediate. What is remarkable is that the letter, like all its forerunners on economic subjects reproduced by Sraffa, concerned monetary issues, while the very next letter reproduced by Sraffa, dated 2nd. March 1814 and addressed to James Mill, like the subsequent ones, concerned the profits of capital and the draft of Ricardo's 1815 essay on profits. (Refer; Sraffa and Dobb, 1951–55, Eds., op. cit., Vol. VI, pp. 100–102). Ricardo's discussions with Malthus and other correspondents switched abruptly and completely from money matters to matters of rent, corn and profits at exactly the time that Ricardo's mind turned to becoming a country gentleman, and of his withdrawal from the Stock Exchange. The single exception was during the period that Ricardo was pestered by Pascoe Grenfell into writing *Economical and Secure Currency* during the summer of 1815. (Refer; Sraffa and Dobb, 1951–55, Eds., op. cit., Vol. VI, pp. 241–305).

In the letters apparently misplaced by Bonar and correctly placed by Sraffa, which were the basis of G. L. S. Tucker's 1954 paper "The Origin of Ricardo's Theory of Profits," (*Economica,* Vol. 21, pp. 320–33.), this remarkable coincidence of personal and discursive factors is also apparent. Tucker suggested that two letters written by Ricardo to Malthus, dated 10th and 17th August 1813, represented the earliest mention of Ricardo's theory of profits, and that, the bullion controversy, Vansittart's recently announced Finance Plan, and/or the Corn Law question raised in parliament in June 1813, rather than the 1815 Corn Law debate, were therefore the foundation of that theory. What is interesting here is that it is not only in terms of the theoretical discussion that these two letters stand out. They stand out also in so far as they express

Ricardo's concern to find a country residence. Ricardo reported that he was during August making "short excursions . . . looking after a house, likely to suit." (Quoted Sraffa and Dobb, 1951–55, Eds., op. cit., Vol. VI, pp. 92–93). Quite apart from the fads and fashions of political debate Ricardo's mind, fleetingly in 1813, was upon country estates and not upon currency. It appears, therefore, that the emergence of the corn model and a new theory of rent had a micro-sociological foundation, at least in part, just as monetary theory had.

121. Sraffa and Dobb, 1951–55, Eds., op. cit., Vol. III, p. 15.

122. N. J. Silberling, 1924, "The Financial and Monetary Policy of Great Britain During the Napoleonic Wars; Part II, Ricardo and The Bullion Report," *Quarterly Journal of Economics,* Volume 38, p. 404.

123. B. Russell, quoted K. W. Rothschild, 1971, Ed., *Power in Economics; Selected Readings* (Harmondsworth: Penguin), p. 7.

CHAPTER 6

The Success of Bullionism and the Bullionists

*We want to look at the thinkers of a given period as representatives of different styles of thought. We want to describe their different ways of looking at things as if they were reflecting the changing outlook of their groups; and by this method we hope to show both the inner unity of a style of thought and the slight variations and modifications which the conceptual apparatus of the whole group must undergo as the group itself shifts its position in society.**

In this final chapter of the case study, attention is turned toward the debate that surrounded the adoption of bullionist monetary policy in the form of the resumption of specie payments by the Bank of England in 1819. The debate spanned the period from 1818 to 1821. In contrast to the earlier phases of debate the focus was by now almost exclusively on matters of policy, rather than principle. Indeed it appears to have been widely accepted during this phase of the debate that the principle of convertibility during peacetime was not, and had never been in question, despite the fact(s) of the continuation of suspension during peace in 1802 and from 1815 onwards.

THE SETTING: 1815–1819

Debate subsided in 1812 with the apparent reconciliation of the City factions, despite the increasing deviation of exchanges from par, and the rising price of gold circa 1813. The next phase of the debate arose in 1816. During the intervening years there were a number of changes in the political and institutional scenery that may be said to have affected the course of subsequent monetary debate.

K. Mannheim, quoted in A. P. Simonds, 1978, *Karl Mannheim's Sociology of Knowledge* (Oxford: Clarendon Press), p. 91.

Table 6.1. Loans for Great Britain and Northern Ireland, 1810-1820

Date of Contract	Sum Raised (million pounds)	Competitive Bid
16 May, 1810	12.0	Yes
20 May, 1811	12.0	Yes
16 June, 1812	22.5	No
9 June, 1813	27.0	No
15 Nov, 1813	22.0	No
13 June, 1814	24.0	No
14 June, 1815	36.0	No
9 June, 1819	12.0	Yes
9 June, 1820	5.0	Yes

Source: P. Sraffa and M. H. Dobb, 1951-55, Eds., *The Works and Correspondence of David Ricardo* (Cambridge: Cambridge University Press), Volume 10, pp. 80-81.

In 1812 there was an extensive outbreak of Luddite insurrection.[1] Spencer Perceval, prime minister and chancellor of the Exchequer, was assassinated in the lobby of the Houses of Parliament on 11th May 1812. Though all the evidence suggests that the assassination was unrelated either to general economic distress or to any wider political plot, it created great concern at the time. Indeed, law and order was a major issue in 1812. Perceval's death also acted to bring about changes in the ministry. Lord Liverpool,[2] son of the first Earl of Liverpool, the former master of the Royal Mint and author of *A Treatise on the Coins of the Realm,* (1805), became prime minister. Addington, "The Doctor," and his close ally Nicholas Vansittart returned to office, the former as secretary of state and the latter as chancellor of the Exchequer.[3]

We have already seen that the factionalism which characterized the City of London financial community appeared to have been suddenly replaced by a spirit of cooperation. The deaths of certain key figures had left room for a more peaceful ascent to dominance for the parvenus—the Stock Exchange community, and other former outsiders. The annual government loans of 1812 through 1815 were shared by the contractors.[4] (See Table 6.1.) This is suggestive of a reconciliation. The end of the huge war loans (June 1816) could be expected to signal a reemergence of factional conflict, and hence of monetary controversy.

The change of ministry effected by Perceval's assassination had a marked impact on trade. The controversial trade restricting Orders of Council, which had been imposed in 1809, were revoked in June 1812.

There were two notable consequences of this. First, the repeal acted to fuel the growing commercial boom as disrupted and curtailed trade resumed. Second, the repeal offered the chance of reconciliation between the incoming Liverpool ministry and the liberal interests who had so opposed the Orders. Economic liberals could now begin to align with the Tory leadership; they were no longer forced into the political liberal camp by their commercial liberalism. This may well have been an important avenue of reconciliation and rapprochement. The cooperative social environment in the City ensured that bills to continue restriction passed through parliament in 1814, 1815 and 1816 with very little debate.[5]

On the economic front the situation that had brought the crisis of depreciation and exchange variation in 1810 was virtually repeated in 1813. Despite the Continental Blockade of 1809, the opening of Brazil and other parts of South and Central America to British trade in 1808 led to a commercial boom during 1809. In this context the Bank of England gave speculation its head and was free in its accommodation of commercial bills, such that there was a rapid expansion of credit in 1809–1810. This was followed by the inevitable collapse and crisis during 1811. Following this short-lived but severe crisis, a recovery through equally easy credit terms at the Bank of England, together with renewed heavy government borrowing, led to another commercial speculative boom. This reached a peak in 1813, when gold and silver reach their maximum prices for the period. (See Table 4.3 above.)

During the period 1814 to 1816 there developed another bout of economic distress, falling prices, and bankruptcies, which was of a similar extent, albeit gentler and more long drawn out than the previous crash. (See Table 6.2.) By 1814 the empire of Napoleon was rapidly contracting back toward the former borders of France, and exporters had gambled that the thus-reopened Europe would be anxious for the goods which had for so long been excluded. The expense of war had been great, however, and few possessed the means to pay for imports.[6] Consequently, the speculative trading quickly collapsed. A bumper harvest in Britain in 1815, together with the grain imports which the previous season's high prices had attracted, produced a glut of grain and a rapid fall in grain prices. This severely affected the country banks because of their agricultural connections.

Bankruptcies, especially bank failures, effected an enormous contraction of credit. According to McCulloch,

> over 200 (out of about 700) Country Banks became altogether bankrupt, or at least stopped payment, in the period 1814–16, and the total dimi-

Table 6.2. Prices and Bankruptcies, 1814-1816

Price falls 1814 to 1816

Gayer Wholesale Index	- More than 20 per cent.
Silberling Wholesale Index	- Almost 30 per cent.

Bankruptcies

Year	Monthly Average
1814	105
1815	147
1816	179
1816 November Peak	227

Source: F. W. Fetter, 1965, *The Development of British Monetary Orthodoxy, 1797-1875* (Cambridge Mass: Harvard University Press), pp. 73-74.

nution of currency in these years was at a minimum 20 millions, probably much more.[7]

One consequence of this crash was a sudden appreciation of the pound. Indeed during 1816–17 the price of gold was only marginally above its mint price, and exchanges with Hamburg were for two years favorable to England.[8] (See Table 4.3 above.) And yet, though resumption had been delayed until the exchanges were favorable, little action toward resumption was taken during 1816–17. Neither the advent of peace nor of favorable exchanges, both proffered as analytical reason, was sufficient to effect resumption.

The wars ended at Waterloo in June 1815. After 23 years of fighting France all but continuously, the enormous effort and psychological burden of the wars were lifted. During the period 1812–15 the scent of victory had forged a new unity in England, but this changed at a stroke in June 1815. In social terms the French no longer united the English from without. People in England began to turn their attention to the social upheavals which had occurred at home: to industrialization, and to changed social relations. The 1815–22 period was a very torrid one for the Liverpool administration,[9] but, where all else had failed, victory united Tories more nearly into a single force—the achievement of the end being a *de facto* endorsement of the means.

The post-war years were marked by considerable economic hardship.[10] Though declining somewhat, high food prices continued after the wars. (See Table 4.2 above.) There was also a marked post-war trade slump as war-related demand subsided. Using 1815 as a base year, exports fell: 1815 = 100, 1817 = 76, and 1823 = 63.[11] Things were slow to pick up after the wars.

Under the leadership of Henry Brougham and Alexander Baring the campaign against property tax (income tax)[12] became the Whig party's political theme for the parliamentary opposition in 1815 and 1816. It was an issue over which the agricultural and industrial interests were understandably, if briefly, almost united.[13] Property tax had been instituted as a wartime measure, and vocal opposition arose when the government announced its intention to halve the rate, rather than remove the tax, at the end of the war. The Whig party-men reveled in the opportunity to take up a cause at once widely popular, and one which would disrupt the government's budgets, hit back at the unpopular and extravagant Prince Regent (George IV), make militarism impossible to maintain, and cast attention on the extent of Tory sinecures. Numerically weak, the parliamentary Whigs made their presence felt in the early parliamentary session of 1816.

The post-war economic distress brought a new wave of popular unrest. The northern hemisphere spring of 1816 witnessed severe rioting. In East Anglia peasants marched under banners inscribed "Bread or Blood," while in 1817 the "Blanketeers" marched.[14] Machine-breaking was also widespread in 1816. Once again the Tory government responded with political repression rather than economic relief. Early in 1817 bills were passed to prevent seditious gatherings, and habeas corpus was suspended.[15] This led, somewhat predictably, to the politicization of the victims. It was in 1816 that William Cobbett first published the two-penny version of his *Political Register,* the "Two-Penny Trash," through which he attempted to direct discontent into the harness of parliamentary reform.[16]

There were also changes on the monetary front. With the exchanges favorable from early 1816, and gold at a premium of barely 1 per cent in October, gold flowed into the Bank of England, and progress toward resumption at the Bank's discretion went ahead during the period from mid-1816 to mid-1817. In October Huskisson called for an immediate resumption of cash payments while the chance was there, and so that credit might be safely *extended* in order to bring economic relief.[17] No official action was taken, but in November the Bank of England began to redeem one and two pound notes dated prior to 1812 in gold coin. This provoked almost no demand for coin. People were apparently accustomed to paper after such a long period, and with no profit to be had from melting coin for export they felt no need for it.

In April 1817 the directors of the Bank of England felt bold enough to offer to redeem pre-1816 one and two pound notes. Again little gold was demanded.[18] By that time, however, things had begun to change. In July 1817 the exchanges again became unfavorable and the price of gold began to rise, such that gold began to flow out of the country. A major contributory impact on the exchanges appears to have been the failure of the 1817 and 1818 harvests, and the consequent need for corn imports. The Bank of England did not heed the warning signs. In September it began to redeem all pre-1817 notes. Over the following few months there was an outflow of gold from Britain worth over 2.5 million pounds.[19]

Following an enquiry relating to the Bank of England and public expenditure in late 1807, a new agreement had been made between the Bank and the government in 1808 regarding management of the public debt. It was subsequently felt that the Bank had managed to wrest a most advantageous deal from Spencer Perceval, one from which it had since derived considerable profit. In 1815 Pascoe Grenfell sought to raise this issue in parliament, and to instigate a reappraisal of those arrangements.[20] There had, since 1800, been a precedent for the government to share in the profits of the operations of the Bank of England according to the stipulations of its charter. Indeed, in 1806 the Bank lent the government 3 million pounds at an advantageous interest rate of 3 per cent until 1814, and a further 3 million pounds interest free in 1808.[21]

Grenfell's resolutions were set down for discussion in the parliamentary session of 1816, and it was in preparation for that event, and for the Bank courts of late 1815, that Grenfell sought Ricardo's assistance in the anti-Bank campaign. He wrote to Ricardo in August 1815 saying that he was "anxious and desirous that [Ricardo] should feel an inclination to undertake . . . to write a short Pamphlet on the Subject to which I have lately called the attention of Parliament."[22] The main motive mentioned by Grenfell was to benefit the public, and to aid the proprietors of Bank stock in participating in the great profits of the Bank of England. Both Grenfell and Ricardo were proprietors of Bank stock.

Grenfell successfully engaged Ricardo, and he showered him with information and documents relating to the Bank. Ricardo wrote the pamphlet *Proposals for an Economical and Secure Currency; With Observations on the Profits of the Bank of England,* and it was published in early February 1816, with a second edition in late February. The work was by no means a simple reflection of what Grenfell sought, and may indeed have been a surprise to him on a number of counts. First, it went much further than Grenfell did in attacking the Bank—even so far as prefiguring Ricardo's later plan for a national bank to relieve the Bank of England of its monopoly of government business.[23] Second, and more importantly, the *plan* for an economical and secure currency was entirely independent of Grenfell's designs.

Ricardo's plan was based on the premise that precious metals were the best available standard by which to maintain a stable value of currency. The main features of his plan to return England and its currency to the par value of 1797 were as follows. First, the Bank of England should pay its cash in bullion rather than coin, so as to avoid the delays and expenses involved in melting and minting. Second, the bullion should only be available in assayed units large enough to make it inconvenient for the public to exchange small-denomination notes. The idea was to dissipate the effects of a panic run upon the banks by avoiding time-wasting, and by placing a bottleneck over the drain. Third, Bank of England notes should be made legal tender, so that country banks might continue to redeem in them rather than in bullion. All these points effectively reduced the level of gold reserves necessary to effect a resumption of specie payments. Fourth, country banks should be required to deposit security reserves with the government as a guarantee of credit, the purpose being to stabilize the country banking situation.[24]

In the House of Commons on 13th February 1816 Grenfell attacked the Bank in the name of public economy on two major counts: first, that the public did not share in the profits derived from the Bank holding public monies, and second, that the Bank made an exorbitantly large claim for the management of the public debt. He went on to argue that the Restriction Act, which he said would have been more aptly entitled, "[a]n Act for relieving the Bank of England from the necessity of paying its obligations in cash,"[25] had allowed the Bank to increase its note issue, and thereby to reap windfall profits—currently some 800,000 pounds per annum. Grenfell found it remarkable that the government agreed in 1806 to pay 90,000 pounds per annum interest on a loan of 3 million pounds at a time when the Bank held public monies of almost 12 million pounds.[26] He wondered why the government had paid the Bank of England 780,000 pounds for the use of its own money. In regard to the management of the public debt, Grenfell revealed that the government was paying the Bank of England some 300,000 pounds per annum.[27] He called for a full enquiry, because the deals between the Bank and the government were obviously too favorable to the Bank.

Though defeated on this occasion Pascoe Grenfell continued to call for enquiries into Bank of England affairs, and into the dealings of the Bank in relation to government. The relationship was deemed to be too close. As it turned out, however, 1817–18 saw a turning point in the relation between Bank and government, and over the next few years they became estranged. There were two issues, which surfaced during the 1816–19 period, that made the Bank of England increasingly unpopular. This led the government to withdraw their formerly unquestioning support. The first was the recurrent issue of Bank profits, and the second that of forgery.

Banks had long, and quite generally, been seen as exploiters of the people. During the post-war years of economic distress the profits that the banks, and especially the Bank of England, had been able to make through government needs and through restriction came to appear unfair. The bankers appeared to have profited from war and public suffering. That the directors of the Bank of England were so reluctant to share the high profits with their stock proprietors, or to even inform the proprietors of their situation, made them many influential enemies—amongst them, of course, Pascoe Grenfell and David Ricardo.

The smoldering animosity felt toward the Bank of England burst out into open conflict over the question of forgery. Prior to restriction in 1797 there had been very few cases of forgery. In the seven years 1790–96 there were no prosecutions. After 1797 the Bank of England began to issue notes of lower denominations and forgery became quite common. Whereas in 1806 approximately 300 forged notes were presented to the Bank, in 1817 nearly 29,000 were presented. Most unpopularly the Bank exacted the death penalty from those convicted. Between 1797 and 1815 there were 257 capital convictions.[28] The Bank of England's willingness to hang the poor did nothing for its public image, nor did the Bank's simple refusal of the forged notes. Many respectable people presented the notes they had accepted in good faith, but they were the losers if the notes turned out to be forged. There were a number of petitions to parliament that sought to make the Bank of England stand the loss, on the grounds that it, not the public, was to blame for issuing such easily forged notes. In conjunction with the charge of profiteering, the issue of forgery plunged the Bank's popularity to an all-time low, and encouraged the government to seek to distance itself from it.[29]

It has been suggested that the Bank of England effected reprisals over the ministry's lack of support on the forgery issue. Following the government's failure to come out in support of the Bank in parliament, in April the Bank refused to renew any more Exchequer bills in May and July 1818, until the government repaid 6 million pounds of its outstanding advances.[30] Such a rebuff was indicative of the breakdown in the relations between the Bank and the ministry. It led to a showdown in April 1819. This division meant the finish of Vansittart's Pittite system of finance. It was indicative of factional disputes amongst the formerly united Bank (merchant) and Tory group(s).

In parliament the Whig leadership, basically Greyite at that stage, was enjoying some increase in power as a new group of prominent liberals emerged. The elections of 1818 left the Liverpool ministry in office, but shook its hold on power. The Whigs had gained a little electorally, but their decisive gain was to prove to be in the shrewd choice of issues on which to stand.[31] Following his success in the matter of the repeal of the property tax in 1816, Brougham was to repeat the triumph on the issues of criminal law, and to a lesser extent on retrenchments. In the 1819 session of parliament the ministry was on the defensive and was not always able to get its way, especially in the Commons.

The time was ripe for a new settlement in the order of things. Public attention was once more drawn to the bullion question, Ricardo's *Economical and Secure Currency* ran to a third edition in March 1819, the government and the Bank were becoming increasingly estranged, and the Vansittart-Herries faction of the Tories was losing ground in the behind-the-scenes politicking of the day to the more liberal Canning-Huskisson faction, as the Tory leadership of Liverpool was gaining ascendancy over that of Castlereagh.

AN ANALYTICAL PREVIEW

The change which brought the Liverpool Tory ministry to office in 1812 affected both political and economic leaders. Lord Liverpool's ministry was much more inclined toward economic liberalism than had been the previous Pittite ministries.[32] Together with the revocation of the Orders of Council, this led to a degree of rapprochement between the Tory leadership and leading economic liberals. The economic liberals were no longer forced to be oppositionists. Meanwhile, in the City financial community the Gentlemen of the Exchange had established their position, and, with newfound cooperation and the enormous increase in the extent of government loans, the factionalism subsided. Soon after the government war loans finished, however, the veneer of cooperation was thrown off, and open conflict re-emerged. The end of the wars turned everyone's attention to the social and political divisions within England, to the momentous changes that had occurred during the previous 23 years, and to the economic hardships which the wars had left in their wake. This set the stage for, and was the context for, an upsurge in monetary debate and controversy in 1816.

It was noted at the end of the last chapter that the elites were beginning to approach one another in socio-cultural space. The old

Figure 6.1. The Relative Positions Approaching (1816-1819)

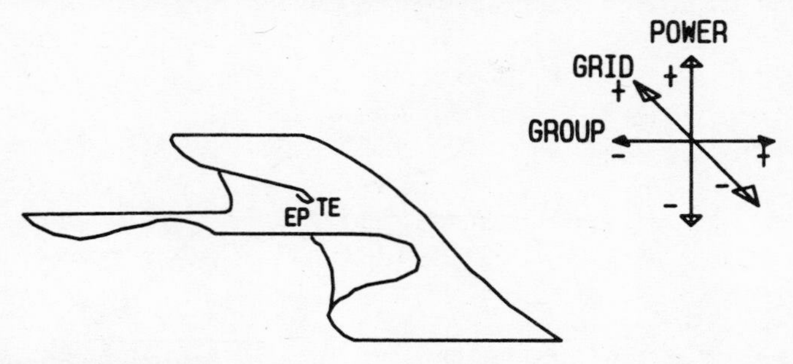

Where TE = The Tory Elite
and EP = The Economic Parvenus

Source: M. Thompson, 1982, 'A Three-Dimensional Model,' In Mary Douglas, Ed., *Essays in The Sociology of Perception* (London: Routledge & Kegan Paul), p.50. Reprinted with permission.

political elite were moving down grid, down power, and, with a lapse in group conflict, down group. The ascendance of the more liberal Liverpool faction broke up the monopoly of the old-guard Tories, and effected a further down grid and down group impulsion. Moving into the ascendancy by expanding their sphere of action, the economic parvenu were, on the other hand, moving slightly down group, up grid and down power. Their power and influence were more diffused, and thus felt less potent. Their orientation began to transcend their local groups, and they lost some within-group regulative power. (See Figure 6.1.)

Figure 6.2 provides the basis for an analysis of what might be expected in the period 1816 to 1821. It entails a projection of the positions of Figure 6.1. The end of the war, and the end of the war loans brought a new scenario, and new forces to bare on the social environment. The economic parvenus were moving down power, up grid, and somewhat down group. They were sliding on a steeply sloping but continuous area of the socio-cultural surface. In this area the fold region is narrower, and they can be expected to have been tending to move back away from the fold. This suggests that both the mode and the content of bullionist analysis would become less extreme.

The political elite, for their part, had begun to move toward a rapprochement with the economic leaders; there was a liberalization beginning. The rise of Liverpool and the decline of Castlereagh, and its effects, is evidence of a decline in group commitment amongst the

Figure 6.2. The Relaive Elite Positions Projected (1816-1821)

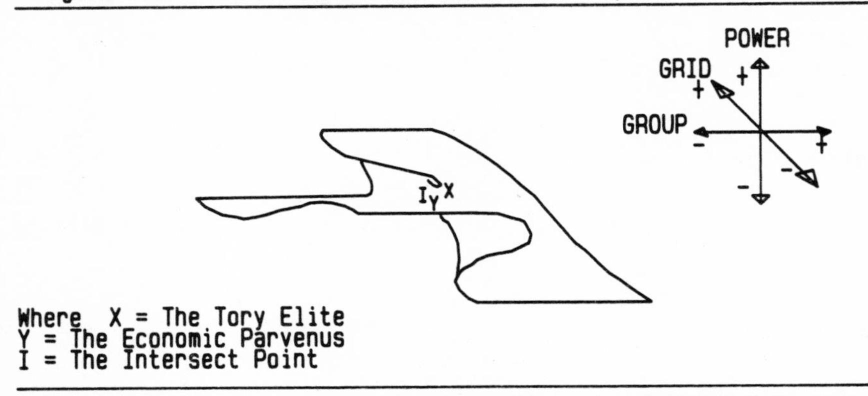

Source: M. Thompson, 1982, 'A Three-Dimensional Model,' In Mary Douglas, Ed., *Essays in The Sociology of Perception* (London: Routledge & Kegan Paul), p.50. Reprinted with permission.

old political elite. They were clearly adopting a strategy different from their former one of ever-increasing exclusiveness. They were, in short, moving quite rapidly down group. These changes led to a dilution of power, which, when added to a more precarious parliamentary position, turned into a diminution of power. Liberalization and weakening rules of exclusiveness also clearly marked a down grid trend.

If these courses, suggested by the socio-cultural schema, are projected, with due allowance for the greater speed of change experienced by the old political elite, they are found to intersect in the area of (I) in Figure 6.2. Interestingly, the transition of both elites to this intersection is via a continuous surface, rather than across a fold (cusp) region. This suggests three things: first, that a resolution of the controversy and conflict, and a widespread acceptance and adoption of a single monetary theory and policy position can be expected; second, that this position will be arrived at through persuasion and reconciliation, rather than sudden revolutionary conversion; and third, the position of the intersection suggests that the compromise reached will be just inside the quadrant occupied by the economic parvenus, and characteristic of individualists. That is, the compromise position reached will be a form of bullionism.

In short, the analysis of the modes of expression and analysis, of the lived-experience of the participants, and of the relative movements and changes within that experience—changes in context, and in the sense of social position—all point to the probability of the success of bullionism, and to the ascendance of bullionist principles and policy. It also points to the fact that the route to this success must lie in the gradual conversion of the old elite. Let us now turn to the details of the debate between 1816 and 1819.

MONETARY DEBATE AND MONETARY POLICY: 1816–1819

The major reviews, notably the *Edinburgh Review* and the *Quarterly Review,* passed quite abruptly from the intense heat of the campaigns of 1811 into silence in 1812. Of a petition from the Lord Mayor and Common Council of the City of London to parliament in 1816, the *Quarterly* reviewers found their apportioning blame to a "delusive paper currency" remarkable. They wrote:

> What! is the ghost of Bullion abroad?—buried as it was "full fathom five" beneath reams of forgotten disquisitions, colder and heavier than any marble monument, what conjuror hath raised it from the grave?[33]

For its part the *Edinburgh Review* was all but silent on the bullion question, until McCulloch's review of Ricardo's pamphlet, *Economical and Secure Currency,* in December 1818. This was, of course, the precursor of the enquiries and the events of the 1819 session of parliament during which, Ricardo and McCulloch agreed, the government had to take "some decided step [to] restore the country to its ancient security."[34]

The *Monthly Review* was also silent on the bullion question. In a lone review of six tracts relating to currency, including two by Lord Lauderdale, in May 1814 the reviewers noted that "[t]he Bullion-question has been destined to exhibit a striking instance of the transient nature of interest."[35] More importantly, in reviewing Lauderdale, the reviewers noted a change which they saw as a consequence of the assassination of Spencer Perceval in 1812. It was, they said,

> unnecessary to notice his Lordship's arguments against a perseverance in the present system [restriction], because the country at large seems to be decided in a wish to depart from it, and the ministry, at least since the elevation of Lord Liverpool, appear to participate cordially in that desire.[36]

That is to say that they noted that the Liverpool ministry, in contrast to the Perceval ministry, appeared to favor resumption. Importantly in our context, Liverpool's ascendance was evidence of the wane of Foreign

Secretary Castlereagh as a leader of the Tories. This led to the resurgence of the Canningites, since Canning's old foe was on the way out.

On 1st May 1818 George Tierney called for the House of Commons to stand by its promise to resume payments at the Bank of England in July. He suggested that if it did not now take a stand restriction would go on year by year, as it had to date. Tierney referred to the principles of the 1810 Bullion Committee in relation to the activities of the Bank—which had extended issue and bought gold through 1817. He suggested that this was not consistent with the Bank's claims to be preparing for resumption.[37] Tierney saw a need for the government to develop and institute a plan which would force the Bank to take appropriate action,[38] since it had shown itself to be unable and/or unwilling to do it itself. To this end Tierney proposed that a committee be appointed. Vansittart, then chancellor of the Exchequer, responded to the motion by ridiculing the practical outcome of the 1810 Bullion Committee, which, he said, "however ingenious and laborious, had led to no practical result."[39] Huskisson replied that

> the House would do well to study the Bullion Report and related publications again, since the Committee's findings had never been answered.[40]

Tierney's motion was lost 164 to 99,[41] but this represented the most support yet experienced for a pro-resumption motion.[42]

Arguments aside, it was by then too late to hold enquiries and pass resumption legislation during that session of parliament.[43] A one-year extension of restriction, until July 1819, was passed, and parliament adjourned in June. It did not reassemble until January 1819, following the tumultuous general election which saw Ricardo secure his seat in the House of Commons.

In anticipation of subsequent parliamentary discussion, and by way of reminding the public, there appeared a review of Ricardo's pamphlet *Economical and Secure Currency* in the *Edinburgh Review* of December of 1818.[44] The review was J. R. McCulloch's second for the *Edinburgh Review,* following that of Ricardo's *Principles of Political Economy and Taxation* in the previous issue.[45] On this occasion McCulloch spent considerable time and effort in supporting Ricardo's plan. He also toed the line of the bullionist cause. "Nothing," concluded McCulloch, "but rendering banknotes exchangeable for cash or bullion, can restore the currency to a sound state."[46]

McCulloch argued against the Real Bills Doctrine. He suggested that the demand for discounts came from the rate of profit and was unrelated to the security of, or the extent of, existing discounts. It was a widely held view at the time that the currency appreciation of 1815–16 demonstrated that the wartime expenses had been the cause of depreciation, contrary to the views of the *Bullion Report* and of bullionist writers. In response to this McCulloch argued that the collapse in profits after the end of the war had reduced the demand for accommodation at 5 percent. Thus, as the country banks failed and no new accommodation was sought, the aggregate quantity of currency was reduced, and the currency appreciated.[47] McCulloch saw this scenario as proof of the bullionist arguments. Once again the contrast between the modes of discourse is apparent. Whereas the anti-bullionists referred to the surface empirical correlation, the bullionists (re)interpreted the events in terms of an underlying causal relation—characteristically hierarchist and individualist styles respectively.

George Tierney, the leader of the Whigs in the House of Commons, had long been launching attacks on the government and the Bank of England: on their relationship, on the system of finance, on the interested motives of the Bank directors, and, most of all, on the continuation of restriction. In an increasingly precarious position after the elections of 1818 the Tories were more than ever vulnerable to such attacks. With the often-promised resumption of specie payments due in July 1819, Grey and the Whig leadership selected Bank restriction as the key focus of the opposition in the 1819 session of parliament.[48] The reasons for this would appear to have had more to do with party political advantage than with the principles of political economy. Grey clearly saw the issue as the best chance of bringing down the Liverpool ministry. He wrote to Lord Holland that it would be impossible "to resume cash payments, without encountering a degree of distress, which no Administration can encounter."[49] That is to say that the Whigs planned to push resumption through in order to oust the Tories in the face of the economic hardship which the measure would cause.

When parliament resumed in January 1819 Prime Minister Liverpool at first tried to avoid the issue of restriction. He then tried to continue as before by introducing a motion to extend restriction for one more year. Four days later, however, he reported to the House of Lords that a communication had been received from the Committee of Bank Directors.[50] The directors of the Bank of England had written to Lord Liverpool, saying that given the present unfavorable state of the exchanges, an extra year would not be sufficient time to prepare for resumption, and that rather than promise the impossible it would be better to submit to a parliamentary enquiry.[51] One might speculate, in the light of later events, that the directors saw this as the best available delaying tactic. In any event Liverpool apparently realized that it would be all but impossible to get parliament to pass *any* further extension of restriction, let alone one of more than a year. With this in mind, and in the context of their worsening relationship, the ministry decided to abandon the Bank.

It was apparent to the ministers that Tierney's motion for a full parliamentary enquiry was dangerous, and that if they were to be defeated by it in the House they would be obliged to resign, or be forced to continue in office in such a weakened condition as to make that office "a positive evil."[52] Consequently, having decided to abandon the Bank of England, the ministers sought to regain the initiative in parliament by effecting an immediate *volte-face,* and themselves calling for an enquiry. Contemporary accounts record the surprise of people close to the ministry, at their about-face on the bullion question. Mr. Ward wrote in a letter to a friend:

> Those that are near the scene of action are no less surprised than yourself at the turn the Bullion question has taken. Canning says that it is the greatest wonder that he has witnessed in the political world.[53]

It would appear that once again, this time on the Tory side of politics, the leading political figures sought party advantage, or the minimization of political damage on the issue. They did not act primarily from either the adherence to, or conversion to principles of political economy. William Cobbett described the situation thus: "Twenty-two years have the Boroughmongers been in a state of stoppage of payments. They cannot remain in that state much longer, without being overthrown."[54] This was their position, and, as Martineau recorded, it was in these party pursuits that the ministers encountered some success. There was

> reason to believe that the bold course taken by the ministers on the bank question did produce something of the effect which Lord Liverpool anticipated, and strengthened them both within the walls of parliament and out of doors.[55]

Having decided on the course of an enquiry, there followed much argument about the terms of reference, and over whether it should be a secret enquiry or not.[56] In the House of Lords, Lord Lauderdale questioned the wording of Liverpool's motion calling for the enquiry, since it appeared to preclude the committee from expressing its opinions.[57] It was to *report* on information only. Liverpool's response made it clear that that was indeed the idea. The ministers had learned from their mistake of 1810, where public enquiry had led to the expression of opinions and recommendations so unpalatable to the ministry.

The argument in the House of Commons over the committee being secret was an extended one. This was due in part to Tierney's obvious annoyance at the ministry's attempt to steal his thunder, and to disarm the opposition's main theme so early in the session. It was also due in part to the technicalities of parliamentary enquiries. Members of public committees were traditionally proposed for appointment, while secret committee members were appointed through ballots. Many saw the ballot system as a facade. Sir Francis Burdett suggested that it would save time if the appearance of the ballot were dispensed with and the ministry simply handed parliament its list of names. Mr. Bennet agreed that it was obvious to the House and the country that the ministers controlled the naming of the committees.[58]

The Tory ministerialists were seeking two things: first, that the enquiries be held in secret; and second, that the ministry be able to stack the committees with a preponderance of ministerial supporters. "[I]t appears, indeed, that, in the Commons, the opposition declined taking any part in the process of nomination," wrote Martineau, who went on to record that, as regards the Commons committee, "of the twenty-one members . . . , as appointed by the ballot, fourteen were ministerialists."[59] (See Table 6.3.) William Cobbett sarcastically remarked, "In order that all may be known, the Committees are to be *selected by the ministers;* and, the members are to be *bound to secrecy.*"[60]

Lord Harrowby was made chairman of the Lords committee and Robert Peel (jnr.) chairman of the Commons committee. They were ministerial choices. Of the former it was said that he was "louder than anybody in saying that he shall regard the resumption of cash payments . . . as the consummation of certain ruin to the country," and of the latter that "he knows little about it."[61] Such was the chairmanship, but leadership was a somewhat different matter. With the deaths of Henry Thornton in 1815 and Francis Horner in 1817, only Huskisson of the leading members of the 1810 Bullion Committee remained. He lost no time in taking the initiative in the examination of the witnesses before the Commons committee. On Sraffa's account Lords Liverpool and Grenville took the lead in the Lords committee, while Canning, Huskisson and Frankland Lewis did so in the Commons committee.[62] This leadership counteracted the ministerial influence in the setting of the agenda and the selection of members, because it involved the Whig Grenville and the leading liberal Tories (basically Canningites).

It has been suggested that the witnesses examined were far from representative. Witnesses whom it was realized objected to resumption, to a gold standard, on *principle* were either shut out or shut up. The leading theoretical heretic, Thomas Attwood,[63] was conspicuous by his absence. Other heretics, such as Thomas Smith, Mathias Attwood, and Hudson Gurney, who did manage to get a hearing, were cut short and were not encouraged to expound their ideas.[64]

Table 6.3. Membership of the 1819 Parliamentary Committees

Lords Committee	Commons Committee
WHIG	WHIG
* Lord Grenville	Tierney
Lord Lauderdale	Lamb
Lord King	Bankes
Marquis of Lansdowne	Mackintosh
	Littleton
	Newport
	Lewis
	Grenfell
	Abercromby
TORY	TORY
Earl St. Germains	* Huskisson
* Earl Liverpool	* Canning
Earl Harrowby	* Frankland Lewis
Viscount Gordon	Vansittart
Viscount Granville	Peel (Jnr.)
Lord Redesdale	Lord Castlereagh
Earl Bathurst	Wellesley Pole
Earl Graham	Robinson
Duke of Wellington	Nicholl
Earl Aberdeen	Wilson
Duke of Montrose	Stuart Wortley
	Manning
	Ashurst

* Leading members.

Source: Hansard, J. 1953. *Hansard's Catalogue and Breviate of Parliamentary Papers, 1696-1834* (Oxford: Blackwell), Volume 39, p. 281, and p. 289.

Once it got about that the ministers were even feigning to consider resumption, a number of petitions began to flow into parliament. On 22nd January 1819 the merchants of Bristol submitted a petition calling for the continuation of restriction to the House of Commons through Hart Davis. By 1st February the merchants of Leeds joined the protest. It is notable on this, and similar subsequent occasions, that there were suggestions in parliament that these petitions were being fabricated by bankers, and were not really representative of public opinion.[65] Consequently, though a steady flow of petitions was presented, they stood somewhat discredited from the outset. As Tierney pointed out, the words chosen by the merchants and manufacturers as far apart as Liverpool, Halifax, and Bristol were so strikingly similar as to raise a degree of scepticism.[66] This undoubtedly played a part in the apparent deafness on the part of the parliamentary committees to the public clamor, and their tendency to dismiss it.

One of the few major sources of opposition to resumption in 1819 was the, so-called, Birmingham economists. The leading pamphleteer of the group, Thomas Attwood, wrote a series of pamphlets and open letters both before and after the committees of enquiry were formed. The twin major aspects of these are by now predictable. The key to Attwood's analysis was an unusually broad definition of money. He suggested that money consisted of "bills of exchange, transfers, book debts, banknotes, gold, silver, and indeed everything that passes for money in any shape or way."[67] This was, of course, in complete contrast to Ricardo's narrow focus on banknotes. The second key aspect of Attwood's pamphlets and letters was his assessment of the broader implications of resumption. He wrote that the Restriction Act had been unjust to the monied interests of the country, but was by 1819 firmly established in a new set of relations "on which all the new debts, and obligations, and establishments of society have been formed."[68]

He was right about the establishment of a new set of relations since 1797, but perhaps failed to appreciate that it was these very relations that were under attack. The Birmingham economists displayed a clear vested interest in Birmingham and what it represented. This was not a cynical money-grabbing exercise, but an artifact of the cultural bias that was the corollary of the way of life in Birmingham. It was the new industrial area that had developed in the context of wartime demand, establishing its historical links to armaments manufacture. In the broader social environment it was the very relations between Birmingham, with its manufacturing, and London, with its government bureaucracy controlled by a landed elite, that underlay the controversy. There were two quite distinct ways of life involved. At the helm in London were a land-based Tory aristocracy and a government hierarchy of long tradition, while growing up in Birmingham were individualist entrepreneurs who sought equality of opportunity for themselves and some say in the economic government of the country.

In addition to this misreading of the climate, the Birmingham economists committed two tactical mistakes in their campaign against resumption. First, they tended to express their views in rather extreme language, in the language of zealots.[69] This may have been due to an element of political naiveté. Second, there was a tendency to pragmatism and/or an excess of ideas which led Attwood, for example, to make numerous, often conflicting suggestions about what should be done.[70] While Ricardo stuck to one plan, Thomas Attwood put forward a plethora of plans. In so doing, Attwood appeared the less convincing.

Another pamphleteer of the period who has been attributed with considerable influence was Reverend Edward Copleston.[71] Copleston published *A Letter to Rt. Hon. Robert Peel, On the pernicious effects of a variable standard of value* on 8th January 1819. It quickly became very popular. It, and its sequel, saw numerous editions, and were widely discussed at the time of the committees' enquiries—so much so that one must suppose its influence to have been considerable. The chief merits of Copleston's first letter appear to have been its very clear exposition of the situation, a basic agreement with bullionist principles, and a very careful assessments of the actual effects of the changes of value—especially in relation to the lower classes.[72] Copleston roundly attacked Vansittart's version of things.

It would appear that Robert Peel found Copleston's letter particularly influential, and it may have been a material factor in his change of heart during the period of the hearings. Copleston, an Oxford fellow, had been Peel's tutor at Oriel College,[73] and correspondence between them over the years suggests that Peel often deferred to Copleston's ideas. In this specific case Peel even went so far as to refer to bullionism as Copleston's principle,[74] although there was really nothing new or original in his analysis—nothing which had not been said in 1810–11.

The parliamentary committees examined witnesses and held hearings from early February until 1st May 1819. The evidence given by David Ricardo and Alexander Baring characterizes the issues involved. Central amongst these was the issue of the deflationary effects of resumption at the old par. Baring had dissented from the report of the Bullion Committee in 1810, and in 1819 he sought to warn of the dire deflationary consequences of resumption. Discussion centered on the extent of the deflation which would necessarily follow resumption, and on ways to minimize it. It is notable that amongst the political leadership and the philosophers, as well as amongst the practical men of business, it was presupposed that deflation to some degree would result. It cannot be said to have come as a surprise to anyone.

That the price of gold would have to be reduced to the mint price of three pounds seventeen shillings and tenpence-halfpenny, before resumption of payments in specie at the 1797 par could long be maintained, was simply a matter of fact. The area of debate that remained was over the question of the extent of the change in the general level of prices necessary to so reduce the price of gold. Would price cuts need to be of a larger per cent than that of gold, or perhaps a lesser percent? In commenting on the evidence presented by William Haldimand, Mrs. Jane Marcet's brother, before the Lords committee, J. L. Mallet recorded in his diary that Haldimand apprehends that no scheme can prevent the extreme distress which will be felt from narrowing the discounts of the Bank. The depreciation is reckoned at about 7 percent. . . . Ricardo does not think that the distress will be so great but in this he differs from all other commercial men.[75]

The deflation issue was overlaid by the even more equivocal issues of the relative price stability of gold vis-à-vis other metals, and the factors which might be affecting that stability.[76] Amongst these factors were the colonial changes and wars of independence in Central and South America which emerged as a consequence of the Napoleonic wars, and the return of many of the affected European countries themselves to metallic standards at that time. Though apparently cognizant of such factors Ricardo made light of them. In evidence before the House of Commons committee Ricardo excused his lack of consideration of real (trade) factors by saying that he was not himself engaged in trade.[77] He maintained throughout that the necessary deflation of general prices was to be measured by the extent of the difference between the mint and the market prices of gold, and would not vary significantly from that amount.[78] As that price differential was then contracting to approximately 4 percent, Ricardo suggested that that would be the extent of the deflation.

Alexander Baring[79] gave evidence to both the Commons and the Lords committees. His arguments acted as the main counter to those of Ricardo.[80] It must be noted that Baring declared himself in favor of the resumption of cash payments in principle, but was very cautious about how and when to achieve that goal. Baring was also in agreement with Ricardo as regards the benefits of a largely paper circulation, and he welcomed Ricardo's plan to institute a graduated resumption of payments in bullion. Nevertheless, from the point of view of the controversy at that stage he and Ricardo can be said to represent opposite sides.

Baring saw that inconvertible currency had permitted a credit expansion that would not otherwise have occurred, and in this he saw some benefit in terms of the expansion of economic activity. In marked contrast to Ricardo's abstract theoretical approach, Baring adopted an historical-institutional approach.[81] He measured inflation/deflation in terms of general commodity price movements, rather than merely the price of gold. What is significant, from the point of view of the later development of economic theory, is that Baring stressed the dynamic, while Ricardo remained at the level of statics. Baring was at pains to stress that the benefit of easy credit was "more to be found in the progression, than in the actual state of things."[82]

These are, of course, the by-now thoroughly expected styles of thought that socio-cultural analysis suggested. Indeed it is notable that Ricardo and Baring agreed on much, and yet they have been seen as the chief opponents in the debate during that phase of the bullion controversy. This shows the centrality of the style of thought, of the mode of analysis, compared with the relative peripherality of the analytical detail and/or empirical content. The contrast of Baring's style of thought with that championed by Ricardo was clear. Indeed so foreign were their approaches that Baring and many others of his ilk continued to find Ricardo's adherence to speculative principles both a danger and a very genuine puzzle.[83] They were speaking a different language. Baring, with a merchant background of long standing, pointed to the complexity of the case and sought a detailed analysis, while Ricardo stuck to a plan to enforce resumption as a matter of principle, based on an abstract analysis of the simple underlying causal relations of political economy.

There was, however, another aspect of Baring's evidence which undermined his theoretical subtleties in the eyes of the policymakers. He agreed with the necessity of a real standard of value for the pound in *principle,* but he stressed that any moves toward resumption must be gradual, taking perhaps 5 to 6 years, and above all should not be undertaken then. Baring's late entry into the committee hearings was due to his being away in Paris on business, and that business was in lending assistance to the French government in its public borrowing. Baring Brothers was the principle organ of very extensive French loans in 1818 and 1819. And, as J. L. Mallet recorded in his diary:

> Narrower means of credits, a closer system of discounts, a return to sound currency in this great commercial country, could not fail affecting Europe for a time; and it is *for a time,* and for that *very time,* that Baring wants facilities of every kind.[84]

Perhaps foolishly, Baring admitted to the House of Commons committee that resumption would interfere with the French loan negotiations.[85] Baring thus appeared personally and conspicuously interested in the continuation of restriction. This could only serve to devalue his input.

Baring disagreed with Ricardo's analysis of the extent of reduction of commodity prices necessary to reduce the price of gold to its 1797 par. He did not think that it would be limited to the extent of the variation of the market above the mint price of gold. He argued that Ricardo was ignoring real factors. He pointed to the impact of resumption in England on the world commodity prices, and the blow to business confidence that tighter credit would cause. In the event, Baring's predictions as to the extent of the deflation consequent on resumption proved to be very close to the mark, while Ricardo's assurances that 4 to 6 per cent deflation was the worst to be feared not only proved to be wrong, but dogged him for the rest of his career.

Latter-day analysts commonly applaud Baring's analysis and decry the simplistic efforts of Ricardo. Gordon wrote:

> This was a case where, given his extraordinary propensity to abstract from the state of expectations or business confidence as meaningful economic variables, Ricardo was ill-equipped (by comparison with the sensitive and wide-ranging Baring) to act as a guide to the monetary realities of his time.[86]

Fetter wrote of Ricardo:

> He virtually ignored the complication that might follow from a further reduction of commodity prices due to other causes being added to the price reduction he anticipated from the reduction in the price of gold. Ricardo's whole attitude gives basis for the belief, held both by his contemporaries and by later generations of economists, that he underestimated the problem.[87]

And yet it was Ricardo's plan that was adopted, and Baring's that was dismissed.

What accounts for the failure of Baring to win his case, and for the triumph of Ricardo? Why did bullionism, given the obvious weaknesses of the arguments of the principal bullionist, succeed in 1819? It is clearly a case in which the weaker theory won out—the theory with less empirical content and which generated less-accurate predictions.

From the perspective of the socio-cultural schema suggested herein this mystery is no mystery at all. The styles of thought are the key, and the victory of one or other of the arguments built upon those styles of thought depends little upon the subtlety or even coherence of the arguments. It depends rather upon the victory of, or progressive ascendance to dominance of, the way of life or set of social relations that accords with the cultural bias upon which the particular style of thought is based.

In the parliamentary enquiries the issues soon evolved into a debate of the technicalities of Ricardo's plan. In its essentials the plan that Ricardo recommended to the committees was the one he had outlined in *Economical and Secure Currency* in 1816, and to which McCulloch had, with an astute sense of timing, drawn attention in the *Edinburgh Review*. Indeed the only difference of detail was that Ricardo now appeared to favor gold rather than silver as the standard. Ricardo made much of his plan in the committee hearings, and, one may assume, behind the scenes.[88] There was not, however, any great degree of support for the ingot plan amongst the witnesses to the committees, except from Alexander Baring and Swinton Holland—one of Baring's partners.[89] This is one more area in which the chief opponents were in substantive agreement, and it highlights the importance of their styles of thought over and above the analytical details.

The plan was, of course, *designed* to minimize the deflationary effects of resumption, but it had other advantages in the power politicking of the day. One great advantage was that it was concrete; it offered the public evidence of action from the first. Thus it was indicative of the government's good faith, and tangible evidence of the sincerity of its resolve to resume payments. Or, perhaps more accurately, it gave the appearance of a resolve.

The reports of the two parliamentary committees, released in May, were substantially the same. In addition to the repayment of 10 million pounds to the Bank of England, the committees recommended gradual resumption, and the repeal of the ancient laws restricting the bullion trade. The plan laid down for resumption followed that recommended by Ricardo but for two points of detail: first, that the ingots exchanged at the Bank were to be assayed and stamped at the Royal Mint; and second, that after 1823 the Bank could revert to specie payments in coin.[90] It was Thomas Tooke who, while agreeing with Ricardo's plan, suggested to the Lords committee that the ingots be issued by the mint rather than the Bank of England. He thought that bullion so issued would "carry more weight" than if it emanated from a private company.[91] It would also, of course, relieve the Bank of England of responsibility and power, and return some of that power to the Royal Mint.

The reports recommended a four-phase slide toward resumption extending from 1st February 1820 to 1st May 1823. It was stipulated that payment in coin would not commence until after 1st May 1823, and only then after one year's notice from parliament.[92] The point of this last was to make it unnecessary for the Bank to build up gold reserves through purchase. The committees followed Ricardo in the wish to prevent such buying of gold in order to minimize the deflationary effects of resumption.

As soon as Lord Harrowby had presented the report to the House of Lords, Earl Grey asked for more time to consider the matter. Lords Liverpool and Landsdowne saw no need for delay and sought the earliest possible consideration, but Lauderdale also spoke out in favor of more time. He said that the members could not possibly digest and give due consideration, in the ten days allowed, to what had taken the committee three months to produce.[93] It is indeed a most marked and notable contrast between the *Bullion Report* and the report(s) of 1819 that in the latter case events were so conspicuously hurried. Liverpool clearly sought to press the matter ahead. He declared himself

> anxious to return to a standard of value; anxious to return to the ancient standard; anxious to return to it with the least practicable delay; anxious to return to it with the least possible distress.[94]

Lord King spoke in favor of the planned resumption. He, perhaps, came to the heart of the issue for the Lords when detailing why he supported the plan, or, perhaps more accurately, in suggesting to the other lords why they should do so. He said that he welcomed the deflationary tendency of the plan because it would favor rentiers at the expense of capitalists. The former had suffered through the years of inflation, but now the balance would be redressed. He suggested that by putting an

> end to Restriction, an alteration in the employment of capital, and the distribution of wealth must be produced in an opposite direction to that which the depreciation of the currency had produced.[95]

King claimed that resumption at the old par would kill off overtrading, implying a moral cleansing. This is surely, in its rudiments, the key element in the thinking that determined the matter of the principles of political economy in the House of Lords.

The second report of the committee of secrecy was presented to the House of Commons by Robert Peel on 6th May 1819. Within his extended speech Peel noted certain aims to which the committee had lent, and for which they had proposed the means of attainment: first, that parliament should exercise greater control over the dealings between the Bank and the government, and that the two should be seen as independent of each other; second, that restriction must continue beyond the current term to avert too great a deflation; third, that the laws restricting trade in gold should be repealed, such that the market price of gold would be a true market price; and fourth, that the best way to obtain the object of a resumption at the old par was the plan suggested. Peel made it clear that they did not see the bullion plan as a permanent measure,[96] nor was there any abstract preference for bullion payment over coin payment—it was merely a practical means to an end.

The committee's recommendations were embodied in resolutions which were moved in the House of Commons by Robert Peel, and debated on 24th May. They provoked a much more extended debate in the Commons than in the Lords, which is interesting in itself. The number of columns in *Hansard's* records provides a telling proxy measure of the relative strengths on the old agricultural and economic parvenus interests in the Houses. Clearly the amount of persuasion necessary to pass a measure which hurt commercial and manufacturing interests while apparently restoring an old score for the landlords was far greater in the Commons than in the Lords. It could, nevertheless, be achieved, since, as Martineau observed of the Commons in 1819, "[t]he commercial and manufacturing interests were most imperfectly represented. The landed aristocracy had retained official power. . . ."[97]

Peel's introductory speech was remarkable. He admitted frankly that he, and by implication many others of the ministerialists, had entered into the enquiry with no such perceptions or intentions as those which they now held.[98] Peel reported that there had been no sign of party spirit in the committee, that he had been converted by what he had heard there, and that

> in consequence of that evidence, and the discussions on it, his opinion with regard to this question had undergone a material change.[99]

Notably, Peel stated that in relation to the *Bullion Report* of 1810–11, he would still vote against the *practical measure,* but now agreed with Horner's *principles.*[100] He agreed with the principles, but not the policy of enacting them in 1811; 1819 was, however, a different matter. By 1819 the argument concerned the questions of *how* and *when.* The *how* was answered by Ricardo's plan, rather than through Bank discretion, and the *when* was to be 1823 (not 1813). Both these points, the institutors and the period of transition, are crucial.

During his speech Peel referred to the tendency for inconvertible currency to encourage overtrading, thereby leading to economic fluctuations which "deranged all the relations of humble life." He went on to say that "[t]he House had too long transferred its powers," and should "recover the authority which it had so long abdicated."[101] In these remarks he touched on the primary points which may be said to have determined the acceptance of bullionism. First, referring to the deranged relations of humble life, Peel noted what all the landholders knew only too well, namely, that the 23 years that had passed since the outbreak of war with France had brought a relatively peaceful revolution in social relations at home. The agricultural interests, the country gentlemen, were losing their hold on power and status, and on the means to achieve them. They were keen, once the realization dawned at the end of the wars, to fight back. Second, parliamentarians sought, as an element of that process, to recover the authority which they had abdicated under the necessity of war to the Bank. They saw a golden opportunity to seize that power once more, or at least to relieve the Bank of England of it. This latter explains why the enthusiasm for resumption was so much greater inside parliament than it was outside; and the former explains why it was more easily passed in the Lords than in the Commons.

On the currency question, as he was to do on Catholic emancipation and the repeal of the Corn Laws, Peel (jnr.)

> surprised the public by suddenly appearing as the chief figure in what we may call the triumph of the principles which up to that moment he had spent his life in opposing.[102]

Walter Bagehot noted this statesmanlike quality in rather more derisive terms. He wrote:

> From a certain peculiarity of intellect and fortune he [Peel] was never in advance of his time. Of almost all the great measures with which his name is associated, he attained great eminence as an opponent before he attained even greater eminence as their advocate. . . . So soon as these same measures, by the progress of time, the striving of understanding, the conversion of receptive minds, became the property of second-class intellects, Sir Robert Peel became possessed of them also.[103]

While amusing, this might miss the point. Bagehot is treating the progress of scientific principles as if it is a purely intellectual process. It is maintained, and it is hoped demonstrated, in this study that this is not in fact the case. Peel was a politician/statesman, and what Bagehot says is no more, or less, than saying that Peel was in tune with the dominant ways of life of his time. Bullionism, the principles of political economy, was no more or less true or valid in 1819 than in 1811, but it was more appropriate to the structure of social relations that emerged after the Napoleonic wars. As Karl Mannheim wrote:

> A theory . . . is wrong if in a given practical situation it uses concepts and categories which, if taken seriously, would prevent man from adjusting himself at that historical stage.[104]

Whereas the campaign against the recommendations of the 1810 *Bullion Report,* mounted and led by a group of backroom figures with close connections with the then-rising Nicholas Vansittart, had been successful in 1811 at the expense of Huskisson, Horner, etc., the tables were now turned. That this happened was closely related to the decline of Castlereagh and the ascendance of Liverpool in the Tory leadership. The Canning-Huskisson faction, so opposed to the Castlereagh Tories who were soon to be back in the ministry, managed in 1819–20 to defeat the Vansittart-Herries faction by means of behind-the-scenes political moves.

It would appear that the moves to gain agreement over the reports amongst the members of the committees involved, in essence, the ministerial members of the two committees. According to Canning's diary, a meeting of this group on 1st April produced an attempt by Lords Harrowby and Bathurst to avoid Huskisson's recommendation of resumption by suggesting a vague plan of their own. Canning's personal diary entries for the next two days reveal something of the play:

> 2nd April . . . Dined with Huskisson, Bining and Arbuthnot. Talked of Bullion for the purposed on impressing Arbuthnot with the necessity of a report on *right principles,* and through him Liverpool.
> 3rd April . . . Meeting at Liverpool's about Bullion. Quite a new proposal. That of Thursday [Apr 1st.] given up. *All* agreed. . . . Is this a consequence of the talk last night?[105]

There was, then, clear evidence of successful political manoeuvring on the part of the Canning-Huskisson faction to gain Prime Minister Liverpool's support for "right principles."

Effectively there was something in the reports for everyone, and thus agreement over them. For the bullionists of principle there were the *right principles;* while for the Vansittart-Herries faction, and many others of the older Tory leadership, there was time. When the parliamentary session began in January, passing an extension of restriction for one more year seemed quite impossible. Yet, by abandoning the Bank of England and Vansittart, and putting their names to abstract principles of which many knew nothing and cared even less, the older Tory leaders were able to put the resumption of cash payments off until 1823—for *four years.* As Hilton observed, undoubtedly correctly for some, the "bullionist professions helped to camouflage the fact that the final consummation was to be delayed longer than had seemed politically possible in February."[106]

When William Manning, a Bank of England director, spoke to the House he was not keen on Ricardo's plan. Manning warned that "[t]he resolutions, if passed in their present shape, would have the effect of fettering the Bank so as to cause an inconvenient reduction of currency."[107] In the context of the failure of relations between the Bank and the government it was, of course, in the eyes of the latter one more point of recommendation for Ricardo's plan that it did fetter the Bank. As Peel said, the House was seeking to recover the authority. . . . In the Lords hearings Ricardo had been asked:

> Is it not also a great Advantage of such a plan, that nearly the whole progress of its operation, and that of our currency as connected with it, would thus be brought successively under the View of Parliament, instead of being left to the Discretion of the Bank.[108]

Many parliamentarians distrusted the Bank directors, or found it convenient to do so. Peel put it diplomatically in parliament, saying that it would be impolitic "to give them [the Bank directors] a discretion as to a plan which they themselves conceived was neither founded in truth nor sanctioned by experience," or to ask them to "act on a principle which they thought untenable."[109] A clear reason for parliament embracing the very principles which they had until so recently rejected was that their *volte-face* on principles allowed them to rationalize and to justify their recovery of power from the Bank of England. This clearly promoted the success of bullionist principles in parliament.

That there was a turn against the Bank in parliament was clear to all. F. W. Fetter noted:

> Two points stand out; first, an increasingly critical attitude toward the Bank of England for determining, under the restriction, policy that properly belonged to parliament, and second, a feeling that, in view of the political unrest and the widespread criticism of the Bank of England as a monopoly fattening on the profits of restriction while it was hanging forgers of its "wretched notes," it was desirable, regardless of the finer points of economic analysis, to settle the money standard.[110]

That is to say that the ministry could at once steal the wind from the sails of the Whigs and strengthen their hold on office, distance themselves from the publicly unpopular, shift the blame for past indulgence and incompetence to a non-government body, (re)claim more power to themselves, and, what is more, delay resumption for three years. Little wonder that they suddenly discovered an unanimity over the principles of political economy.

AN ANALYTICAL SUMMARY

It can be seen that a leading influence in the cause of conversion came from the most liberal of the Tories. These were the men whose position placed them in the forefront of the trend movement of the political elite across socio-cultural space to a new position in the individualist region of the model, the same quadrant that the economic parvenus occupied. Their efforts at the micro level of group dynamics, and in individual politicking, acted to break down the barriers of the old elite group, and to thereby weaken it. The Tory elite were indeed forced to give ground in order to survive in office, in order to retain some hold on power. The economic parvenus had gained too much ground by then to be defeated and denied. It must have been apparent to most members of both elites, once the dust had settled over the jockeying for position in the new context of peacetime, that a new settlement in the order of things was necessary. It merely remained for men to construct the arguments in that new context which might justify their emergent world view to themselves and to others, that is, to reach a compromise theoretical position which made sense of the new context. Again it can be seen that the key players, those at the center of the drive toward the adoption of bullionist principles and policy, were those who were spatially central in socio-cultural space. It was they, broadly speaking the liberal Canningites, who led the Tories to a new position.

In the years after the enactment of Peel's Bill these central figures, the Canningite liberal Tories, continued to be the most influential. It was they who presided over a profound economic liberalization which was embodied above all in the moves toward free trade during the 1820s, with which the names of Huskisson and Robinson are notably attached. This is one more indication of the insights which position in socio-cultural space can suggest. Like Ricardo and Baring, these men were *central,* spatially, analytically and practical politically.

In summary, then, it can be seen how the apparent acceptance of bullionist principles was based in the emergent post-war social relations. Once again it is in the process(es) of group dynamics that we can find the development of, and choice of ideas and actions. At the macro level we have identified the traditional Tory elite and their mercantile and bureaucratic allies with the styles of life and thought characteristic of hierarchists, certain radical utopian reformers with the styles of life and thought characteristic of egalitarians, and members of the emergent financial and commercial elite with the styles of life and thought characteristic of individualists. Each way of life has its own cultural bias affecting perception and choice. At the micro level we have traced the many factional rivalries, and witnessed their influence on the analytical structure and content of the schemas and theories developed and deployed by the actors. We have seen, both in detailed and in general terms, how and why bullionists principles and policy eventually triumphed in the bullion controversy of 1797 to 1821.

Historiographic approaches that ignore these processes and look only to rational reconstruction are inadequate to the task of unveiling the historical dynamic process of the development and diffusion of knowledge. Nor is it sufficient to look only at the means without considering the ends. It is not just a story of a structure of relations, of social networks, of institutions, etc.; it is also necessary to analyze what is taken up and what not. Opportunities, mere possibilities, technical, economic, etc., are not explanatory. We must seek to know which are taken up, how successfully, and why only those taken up are chosen. We must, in short, engage an analysis of the social relations of power which involves an endogenous treatment of change, of choice or selection.

Interestingly, the projection as to the spatial position representing these new theoretical and policy positions points to a relatively steeply sloping region of the socio-cultural surface, (see Figure 6.2 above), the implication being that it was not a very stable position. That is to say that the consensus reached in 1819–21 was rather brittle, and a resurgence of controversy would be expected. This doubtlessly leads us to the, so-called, banking controversy which led up to the far-reaching Banking Act of 1844. But that is another story.

NOTES

1. M. I. Thomis, 1972, *The Luddites; Machine Breaking in Regency England* (New York: Schocken).
2. Robert Banks Jenkinson (1770–1828). Second Earl Liverpool 1808.
3. A. W. Acworth, 1925, *The Financial Reconstruction of England 1815–1822* (London: King and Sons), p. 12. Vansittart had been joint Secretary at the Treasury under Pitt and Addington, but had been out of office since 1807.
4. P. Sraffa and M. H. Dobb, 1951–55, Eds., *The Works and Correspondence of David Ricardo* (Cambridge: Cambridge University Press), Vol. X, pp. 80–81.
5. F. W. Fetter, 1965a, *The Development of British Monetary Orthodoxy, 1797–1875* (Cambridge, Mass: Harvard University Press), p. 62.
6. Acworth, 1925, op. cit., p. 72.
7. Ibid., p. 72, fn. 6.
8. N. J. Silberling, 1919, "The British Financial Experience, 1790–1830," *The Review of Economic Statistics,* Vol. 1, p. 287.
9. J. E. Cookson, 1975, *Lord Liverpool's Administration; The Crucial Years 1815–1822* (Hamdon, Conn.: Archon Books).
10. G. Wallas, 1898, *The Life of Francis Place, 1771–1854* (London: Longmans Green), p. 114, and L. Stephen, 1900, *The English Utilitarians* (London: Duckworth), 3 Vols., Vol. I, p. 217. See also G. F. Langer, 1987, *The Coming Age of Political Economy, 1815–1825* (Westport CT: Greenwood Press).

11. G. Routh, 1975, *The Origin of Economic Ideas* (London: MacMillan), p. 135.
12. Addington had tried, not altogether successfully, to re-christen Pitt's Income Tax to make it more palatable.
13. Harriet Martineau, 1877–78, *A History of the Thirty Years' Peace* (London: George Bell and Sons), 4 Vols., Vol. I, pp. 30–31.
14. J. L. and Barbara Hammond, 1978, *The Village Labourer* (London: Longman), p. 124. And G. D. H. Cole, 1971, *The Life of William Cobbett* (Westport, Conn.: Greenwood Press), pp. 216–17.
15. G. S. L. Tucker, 1976, *William Huskisson Essays of Political Economy* (Canberra, Australia: Australian National University), p. 318.
16. Harriet Martineau, 1877–88, op. cit., Vol. I, p. 74.
17. B. Hilton, 1977, *Corn, Cash and Commerce; The Economic Policies of the Tory Governments 1815–1830* (Oxford: Oxford University Press), p. 39.
18. Acworth, 1925, op. cit., p. 73.
19. B. Gordon, 1976, *Political Economy in Parliament 1819–1823* (London: MacMillan), pp. 27–28.
20. Sraffa and Dobb, 1951–55, Eds., op. cit., Vol. IV, pp. 136–37.
21. Ibid., p. 88.
22. Ibid., p. 242.
23. Ibid., p. 45.
24. Ibid., pp. 54–73.
25. *Hansard's*, Vol. XXXII, p. 463.
26. Ibid., p. 470.
27. Ibid., p. 487.
28. Fetter, 1965a, op. cit., p. 71.
29. Parenthetically, the forgery problem engendered an argument for resumption. The reasoning being that, given that the paper pound was so easily forged, the re-introduction of gold coin would in part solve the problem. Some people began to seek resumption as a means of returning to a coin circulation. The forgery issue also favored the cause of the country banks, because country bank notes were, typically, far less easily, and far more rarely forged than were Bank of England notes.
30. Hilton, 1977, op. cit., p. 35.
31. Harriet Martineau, 1877–88, op. cit., Vol. I, p. 25ff.
32. Cookson, 1975, op. cit., p. 396.
33. R. Southey, 1816, "On Parliamentary Reform," *Quarterly Review*, Vol. 16, No. XXXI, Art. XI, p. 260.
34. T. R. Malthus, 1818, *Edinburgh Review*, Vol. XXXI, No. LXI, Art. III, p. 77.
35. *Monthly Review*, 1814, Vol. LXXIV, Arts. VII–XII, p. 171.
36. Ibid., pp. 182–83.
37. *Hansard's*, Vol. XXXVIII, pp. 443–51.
38. Ibid., p. 457.
39. Ibid., p. 463.
40. Ibid., p. 486.

41. Ibid., p. 498.
42. Fetter, 1965a, op. cit., p. 84.
43. *Hansard's,* Vol. XXXVIII, p. 765.
44. J. R. McCulloch, 1818, *Edinburgh Review,* Vol. 31, No. LXI, Art. III, pp. 53–80.
45. F. W. Fetter, 1953,"The Authorship of the Economic Articles in the Edinburgh Review 1802–47," *Journal of Political Economy,* Vol. 61, p. 249.
46. McCulloch, 1818, op. cit., p. 80.
47. Ibid., p. 64.
48. Hilton, 1977, op. cit., p. 40.
49. Quoted Hilton, 1977, op. cit., p. 40.
50. *Hansard's,* Vol. XXXIX, p. 78.
51. Fetter, 1965a, op. cit., p. 85.
52. Hilton, 1977, op. cit., pp. 40–41.
53. Quoted Harriet Martineau, 1877–78, op. cit., Vol. I, p. 284.
54. Cobbett, 1819, *Political Register,* Vol. XXXV, No. III, pp. 67–68.
55. Harriet Martineau, 1877–78, op. cit., Vol. I, pp. 286–87.
56. *Hansard's,* Vol. XXXIX, pp. 131–33.
57. Ibid., p. 204.
58. Ibid., p. 281.
59. Harriet Martineau, 1877–78, op. cit., Vol. I, p. 271.
60. Cobbett, 1819, *Political Register,* Vol. XXXV, No. II, p. 56.
61. Quoted Hilton, 1977, op. cit., p. 42.
62. Sraffa and Dobb, 1951–55, Eds., op. cit., Vol. V, pp. 353–54.
63. Refer to: S. B. Checkland, 1948, "The Birmingham Economists," *Economic History Review,* Vol. 1, No. 1, pp. 1–19.; A. Briggs, 1948, "Thomas Attwood and the Economic Background to the Birmingham Political Union," *Cambridge Historical Journal,* Vol. 9, No. 2, pp. 190–216.; F. W. Fetter, Ed., 1964, *Selected Economic Writings of Thomas Attwood* (London: London University Press).
64. Hilton, 1977, op. cit., pp. 43–44.
65. *Hansard's,* Vol. XXXIX, pp. 188–90, pp. 212–13, p. 214, p. 276, etc.
66. Ibid., p. 214.
67. T. Attwood, 1817, *Prosperity Restored; or Reflections on the Cause of the Public Distress, and on the only means of Relieving them* (Birmingham), quoted D. J. Moss, 1981, "Banknotes versus Gold; the monetary theory of Thomas Attwood in his early writings, 1816–19," *History Of Political Economy,* Vol. 13, No. 1, p. 21.
68. T. Attwood, 1819, *A letter to the Earl of Liverpool, on the reports of the Committees of the two Houses of Parliament on the question of the Bank Restriction Act* (Birmingham), quoted Moss, 1981, op. cit., p. 34.
69. D. J. Moss, 1981, "Banknotes versus Gold; the monetary theory of Thomas Attwood in his early writings, 1816–19," *History Of Political Economy,* Vol. 13, No. 1, p. 19.
70. Ibid., p. 32.

71. Refer to: Hilton, 1977, op. cit.; Gordon, 1976, op. cit.,; S. Rashid, 1983, "Edward Copleston, Robert Peel, and Cash Payments." *History of Political Economy,* Vol. 15, No. 2, pp. 249–59; etc.

72. S. Rashid, 1983, "Edward Copleston, Robert Peel, and Cash Payments," *History of Political Economy,* Vol. 15, No. 2, pp. 249–59.

73. Ibid., p. 249.

74. Ibid., p. 254.

75. Sraffa and Dobb, 1951–55, Eds., op. cit., Vol. V, p. 353.

76. Fetter, 1965a, op. cit., p. 89.

77. Sraffa and Dobb, 1951–55, Eds., op. cit., Vol. V, pp. 384–85.

78. Ibid., p. 385, & pp. 416–18.

79. Alexander Baring (1774–1848). Later Lord Ashburton. Baring was the leading figure in the London merchant community in the early nineteenth century, and the main force in Baring Brothers Bankers. He became a large scale landowner and served actively in both houses of parliament. He was a director of the Bank of England in 1811. See J. Clapham, 1977, *The Bank of England: A History* (Cambridge: Cambridge University Press), 2 Vols., Vol. 1, p. 31.

80. B. Gordon, 1987, "Alexander Baring versus David Ricardo; Economic Policy and Parliament after 1815," (Newcastle, Australia: Newcastle University), Occasional Paper 132.

81. Ibid., p. 26.

82. Baring, quoted Gordon, 1976, op. cit., p. 37.

83. See, for example, Gordon, 1976, op. cit., p. 96, p. 106, pp. 137–38, p. 147, p. 158, p. 159, etc.

84. Quoted in Sraffa and Dobb, 1951–55, Eds., op. cit., Vol. V, p. 352.

85. Gordon, 1976, op. cit., p. 195.

86. Gordon, 1976, op. cit., p. 39.

87. Fetter, 1965a, op. cit., p. 91.

88. Sraffa and Dobb, 1951–55, Eds., op. cit., Vol. V, pp. 352–53, & p. 356.

89. Fetter, 1965a, op. cit., p. 92.

90. Sraffa and Dobb, 1951–55, Eds., op. cit., Vol. VIII, p. 26.

91. Sraffa and Dobb, 1951–55, Eds., op. cit., Vol. V, pp. 362–63. See also A. Aron, 1989, "The Early Tooke and Ricardo: a political alliance and first signs of theoretical disagreements," *History of Political Economy,* Vol. 21, No. 1, pp. 1–14.

92. Fetter, 1965a, op. cit., p. 93.

93. *Hansard's,* Vol. XL, pp. 224–26.

94. Ibid., p. 610.

95. Ibid., p. 640.

96. Ibid., pp. 152–75.

97. Harriet Martineau, 1877–78, Vol. I, op. cit., p. 47.

98. Ibid., p. 273.

99. *Hansard's,* Vol. XL, p. 677.

100. Ibid., p. 678.

101. Ibid., p. 689.

102. Harriet Martineau, 1877–78, op. cit., Vol. I, p. 274.
103. W. Bagehot, quoted Gordon, 1976, op. cit., p. 198, fn. 16.
104. K. Mannheim, quoted P. Hamilton, 1974, *Knowledge and Social Structure; An introduction to the Classical Argument in the Sociology of Knowledge* (London: Routledge and Kegan Paul), p. 103.
105. G. Canning, quoted Hilton, 1977, op. cit., p. 44.
106. Hilton, 1977, op. cit., p. 45.
107. *Hansard's,* Vol. XL, pp. 740–42.
108. Quoted Sraffa and Dobb, 1951–55, Eds., op. cit., Vol. V, p. 443.
109. *Hansard's,* Vol. XL, pp. 684–85, and p. 714.
110. Fetter, 1965a, op. cit., p. 95.

CHAPTER 7

Summary, Assessment, and Conclusions

*True knowledge, freed from the ideological obscuration of ahistorical "bourgeois" conceptual systems, is thus made possible when the "isolated facts" of social life are integrated into a totality, a knowledge based on a "conceptual reproduction of [social] reality."**

A mainstream of the conventional approaches to the history of science, history of thought or ideas is derived from philosophies of science that are based on classical empiricism. The key features of empiricism are its focus on events as isolated, discrete things, and its identification of causal relations in the constant conjunction of these events.[1] Such an approach has been common in the history of economic thought.[2] In the context of the bullion controversy such an approach would imply that monetary crisis, in the form of depreciation and/or exchange variation, would provoke debate on monetary questions, and thereby be the "cause" of developments in monetary theory. In analyzing the bullion controversy in detail, however, we find that debate was not linked in any direct way to depreciation and/or exchange variation. In the cases of both the Irish currency debate and the English currency debate, and perhaps more significantly, silence on both topics, we find that there is no such conjunction of events. Debate and event are not conjoined.

We find that the Irish currency debate spanned 1803 and into 1804, whereas the maximum depreciation of the Irish pound against gold, and the greatest exchange depreciation of the Irish pound, occurred in 1816. In 1816 there was no debate on Irish currency. Similarly, the

G. Lukacs, quoted, P. Hamilton. 1974. *Knowledge and Social Structure: An Introduction to the Classical Argument in the Sociology of Knowledge* (London: Routledge and Kegan Paul). p. 41.

peaks of the debate on English currency where 1909–10 and 1819, whereas the maximum depreciation and exchange variation of the English pound occurred in 1813. Moreover, we found that there was a quite remarkable silence on the subject in 1813.

Historiography founded on the event ontology of empiricism has, in the process of exploring the bullion controversy, been shown to be weak. The approach based in cultural theory that we developed and applied herein, however, can be shown to have performed far better. Cultural theory reveals the links between the specific timing of debate, and of silence in the debate, with the underlying social relations of the key agents: actors and groups. In this particular case it emerges that it is shifts in group power in the City of London, group conflict and factionalism that provide the key motivations, and the key dimension of change.

The Irish currency debate circa 1803 is revealed to have been motivated by, and underwritten by, the pursuit of group power relativities after the Act of Union with Ireland. Once Irish officials' salaries were paid at Irish parity, rather than English parity, such that the equity between landlords' remittances and officials' salaries had been re-established, debate ceased—despite worsening depreciation. It should be noted that the content of analysis was affected by these motivations because the focus of analysis in the case of the Irish currency debate circa 1803 was on paper parity, and not on the gold value of the Irish pound. It was an analytical question of paper parity, because it was a real social question of elite groups power balance or parity.

In the case of English currency debate, likewise, we find that the power relations of the key social groups underpinned the timing of both debate and non-debate. We found that the height of the English currency debate coincided with the failure of the Gentlemen of the Exchange to win the government loan contract of 1808, conflict over the subscription room status of the Stock Exchange and eligibility of merchant and banker members circa 1809–10, scandal over Abraham Goldsmid's role in the underhanded release of Exchequer bills in March 1810, the deaths of the Goldsmids in 1808 and 1810, and of Sir Francis Baring in 1810, together with a series of commercial failures in 1810, and, of course, with the falling to discount of loan stock. This war in finance is shown to be the real underlying context of monetary debate, a war in the City of London that was being played out behind the war of finance, the Napoleonic wars.

We also found that there was a quite marked lapse in the controversy over English currency between 1811 and 1816, despite the fact that the maximum exchange variation and depreciation was in 1813. Our cultural theory approach unearthed the social basis of this silence. We found that the deaths of Francis Baring and Abraham Goldsmid in 1810 effected a *de facto* change in the balance of power in the City, and allowed the Gentlemen of the Exchange, the parvenus, to gain ascendance. There were cooperative government loan contracts from 1812 through 1815. Once Ricardo and his group achieved ascendance, debate stopped. Ricardo's monetary publication's ended with *The High Price of Bullion* in April 1811, not to recommence until *Proposals for an Economical and Secure Currency* in February 1816. This lapse in publication occurred despite the fact that Ricardo continued to address monetary questions in his private correspondence throughout this period.[3]

By focusing on these phases of the bullion controversy we reveal that conventional empiricist historiography fails. More importantly we show that cultural theory, and the grid-group-power approach derived from it, succeeds. It succeeds in locating the basis of thought in the real underlying social relations of the time. It reveals the inter-relation between social relations and styles of thought, between participation and cultural bias.

More specific philosophies of science that highlight the importance of the logical structure of arguments have also been pressed into service as underpinnings for historiography.[4] In the context of our review of the bullion controversy, there are, perhaps, three aspects that should be considered in the assessment of these approaches. They are the strong logical structure suggestions that that which explains (the explanans) logically entail that which requires explanation (the explanandum), that the most resilient and/or well confirmed hypothesis wins over weaker theories, and that the theory that generates more accurate prediction will triumph. Let us look at these in turn.

The adherence to logical structure implies, for historiography, that there would be a logically necessary relationship between the analysis and the conclusion(s) of an argument that claims scientific status and/or is a part of scientific discourse. Our review of the bullion controversy reveals a number of instances when a dissociation of analysis and conclusions occurred.

In the early phase of the currency debate we observe the disparity between the two chief bullionists as regards the identification of the cause of depreciation and exchange variation. Walter Boyd constructed a characteristically bullionist argument, and concluded that the Bank of England had overissued. John Wheatley constructed an equally characteristically bullionist argument, deduced monetary expansion, and yet concluded that it was due to unregulated country banks, not overissue at the Bank of England. Their analyses in point of style and of content were the same, but their conclusions different. Our cultural theory-based analysis is able to explain both the bullionism of Boyd and Wheatley, their mode of argumentation or style of thought, and that they conclude differently. It emerges that Boyd was a central figure in the City of London, while Wheatley was an outsider; Boyd a part of the key underlying social conflicts, and Wheatley not. Boyd's social relations put him in a specific relation to the Tory elite and to the Bank of England, a relation reflected in that expressed in his monetary analysis.

During the height of the controversy, similarly, we see that Ricardo, Mushet and other key bullionists concluded that the Bank of England had overissued paper, while Chalmers, William Cobbett and the anonymous continental merchant of the Bullion Committee enquiries, Mr. ****, concluded otherwise. In evidence to the Bullion Committees Mr. **** provided a third way in which to link analysis and conclusions, a way that was at once bullionist and anti-bullionist. In essence Mr. **** suggested that trade disruption was the cause of depreciation (anti-bullionist analysis), but that the resumption of a gold standard was the only cure (bullionist conclusion). Chalmers, by contrast, blamed the scarcity of bullion and the issue of notes by country banks, and concluded that trade expansion—increased exports—was the only true restorative. For his part Cobbett focused on commodity price rises, parliamentary privilege and the Tory sinking fund in causal analysis, and concluded that resumption was the solution. In these cases we can see that certain players dissociated analysis and conclusions in ways that the mainstream bullionists and anti-bullionists did not.

The key to understanding these cases of dissociation of analysis and conclusions is not to be found in the logical structure of statements, but rather, as cultural theory shows us, in the underlying social relations of the key players. Specifically it should be noted that Cobbett and Mr. **** were outsiders, outsiders to City of London factionalism. Hence, it is revealed that the City-based factionalism, rather than logical necessity, underpinned the emergent links between analysis and conclusions. Cobbett's analysis was that of an egalitarian; it matched neither that of the individualist style of the bullionists nor the hierarchist style of the Tory/merchant anti-bullionists.

Without wishing to labor the point, our review of the bullion controversy shows that it was not necessary to blame the Bank of England for overissue, and, as a corollary, it was not necessary to deny overissue to excuse the Bank of England. The principles of bullionism could have been independent of any perception of the role of the Bank of England, but they were not. Anti-Bank of England sentiment, opposition to the merchant/banker grouping—their style of life and their mode of thought—went hand in glove with bullionist thought. It was a part of the lives of the key bullionists. The bullionist and anti-bullionist schemas were structured from different, yet specific cultural biases, and reflected different, yet specific structures of social relations.

The second avenue of approach to historiography that traditional empiricist philosophies of science imply is that the stronger theory wins out in a competition between the hypotheses that different theories generate.[5] Were such an approach correct, our history should reveal instances wherein the stronger theory, be it in terms of empirical scope and content or of predictive capacity, emerged and triumphed over weaker ones. Our review of the bullion controversy, however, does not show this.

On the contrary, our review of the bullion controversy reveals two major instances in which a weaker theory was accepted over a stronger one; despite the clear availability of a stronger alternative and ample evidence of its superiority. The first was in 1810 when the bankers' Real Bills Doctrine was accepted as definitive. The bankers of the Bank of England held that the accommodation of bills based on real exchanges of goods could not be inflationary. And yet it was obvious that practice based on this principle had critically depleted the Bank of England specie reserves and effected a depreciation of the pound—in the circumstances of inconvertibility—long before 1810. The Real Bills Doctrine was accepted over bullionism as an explanation as well as a policy practice in 1810, despite an available alternative and contrary evidence.

The second instance of the weaker theory triumphing occurred in 1819, when Ricardo's bullionism, founded on a simple quantity theory of money, was accepted instead of more subtle forms that embraced the velocity of circulation element and/or that did not require such a clear distinction between the real and money economies. There was ample evidence that Ricardo's bullionism was seen at that time as simplistic and unrealistic, and yet it was preferred in 1819. Indeed, despite its conspicuous practical failure in the 1819–21 period in regard to predicting the extent of the deflationary impact of resumption, Ricardo's analysis was accepted as a foundation to economic theory until the middle of this century.

An explanation of these instances has resulted from the application of cultural theory to the bullion controversy. Our approach shows that the Real Bills Doctrine was an expression of commitment to a traditional set of rules, rules that had governed banking but that were wholly inappropriate for a central bank when there was no automatic standard (convertibility). It triumphed as a doctrine, because it was the doctrine of the group who triumphed. Likewise, we uncovered the basis of Ricardo's overlooking the velocity of circulation element, namely, its attachment to Henry Thornton, who was seen at the time as a pro-restriction writer, an anti-bullionist. Ricardo and others bypassed the idea because it was not of their group, not an integral part of their style of thought.

The pro-Ricardian bullionism choice in 1819 leads us to the third area of concern, namely, to the issue of predictive content. It has been suggested by philosophers of science that the theory with the greater ability to predict, which generates hypotheses that are confirmed and/or not refuted by events, will win out. For historiography this means that we should see cases where predictive performance will be a chief test of theoretical schemes. Our review of the bullion controversy, however, suggests that this is not the case.

A key phase of the bullion controversy was that in which the principles of bullionism came to be adopted, and to provide the platform for policy practice circa 1819. The evidence given before the parliamentary committees of 1819 by David Ricardo and Alexander Baring encapsulates the issues at hand. Ricardo deployed a simplistic monetary argument, whereas Baring expressed a greater concern for the effects of real or trade relationships, and for the extent of deflation necessary to bring paper currency back to its specie par. No one could agree with Ricardo's prediction of a somewhat less than 4 per cent deflation. In the event, of course, Ricardo was proved wrong and Baring very close to the mark. Fetter, in a manner similar to a number of others, records that contemporary and later generations of economists saw Ricardo's analysis as both too simplistic and factually wrong.[6] And yet, not only did Ricardo win the argument in 1819, but the principles of bullionism locked into Ricardo's simplistic quantity theory of money reigned until World War II. The "mistake" of 1819 was repeated, with the same theoretical justification, when the gold standard was resumed in Britain after World War I, and, of course, with the same crippling result. The theory that predicted best did not win when Ricardo's bullionism won in 1819. Indeed the opposite was the case, and, as we have shown, it was known by contemporaries to have been the case.

What our cultural theory approach shows is that this case of theory choice was based in the underlying social relations, relations that centered, in this case, on the City of London. Ricardo's analysis was accepted for two key reasons: first, it was in the mode or style of the analysis of the ascending group/class; and second, it was accepted because it came with a practicable plan that suited the purposes of the groups involved. It was a plan that punished the Bank of England, and restored a balance of power in the City and in the nation as a whole.

In all these cases, cultural theory shows that the theories, the content of the theories, and the analytical detail of the theories were linked to factional City plays, to the war in finance. The details of theory construction and instances of theory choice are revealed to have been neither factually based nor logically necessary, but rather contingent choices or options—optional ways of seeing things. They were, in short, perspectives related to their underlying forms of social organization and practice, perspectives that were intimately related to the underlying need for the conceptual justification of these practices, perspectives that were socially constructed symbolic presentations of the social relations of the participants.

In the competition of scientific research programs the one with the greatest empirical content could be expected to triumph.[7] The historiographic implications are clear, but does this describe the bullion controversy? In our review of the controversy we find that it does not. It is clear that the victory of Ricardo's bullionism in 1819 was a case where the theory with the greatest empirical content did not win. But this is not the only instance. Our review of the bullion controversy reveals that in 1810, as soon as Ricardo entered the debate with his abstract form of bullionism, others were building arguments that achieved greater depth of empirical analysis. The reception of Bosanquet is but one example; the cases of Henry Thornton and William Blake are also notable. In the words of Ellis and Canning:

> Mr. Bosanquet presents himself as one of the most formidable champions against the Bullion Committee, and professes to fight them not with arguments but with facts.[8]

Clearly, Bosanquet's approach involved greater empirical content. Similarly, Henry Thornton provided a most detailed and penetrating analysis, as did William Blake, and yet neither were key players.

In the event of the consummation of bullionism as policy practice in 1819, the performance of Ricardo was markedly inferior to that of Alexander Baring. Baring's has been described as a sensitive and wide-ranging analysis, and as historical-institutional, while that of Ricardo has been described as ill-equipped, simplistic, and unrealistic.[9] It is not only latter-day historians who so judge the situation, as our review revealed that this was a common contemporary perception.

A Kuhnian-inspired approach to historiography would suggest that a theory or program that was in crisis, one with a number of anomalies, would be ripe for replacement, and that such a theory would be defeated by a better theory. Our review of the bullion controversy, however, suggests that this was not the case. It has been noted that Ricardo's bullionism, in the form in which it was accepted in 1819, embodied a simple quantity theory of money. An alternative that was analytically and empirically superior existed in the form of the velocity of circulation argument introduced by Thornton, but was not taken up. The deflationary effects of resumption in 1819–21 proved Ricardo's simple bullionism wrong and Baring's institutional analysis right. This deflation provided the kind of anomaly of explanation, of detail of analysis, that should have precipitated a crisis for Ricardian monetary theory. In the event, however, Ricardian theory went from strength to strength, and became the dominant paradigm for at least the next 50 years.

Moving beyond this pattern of analyzing specific historiographic approaches and performances to more general comments, the following points are noteworthy. Our story casts serious doubt on the viability of any "growth of knowledge" notion. An incremental or even cyclical, punctuated growth does not fit the case of the bullion controversy. The triumph of Ricardo in 1819 is widely perceived to have been a retrograde step—retrograde in terms of the quantity theory of money, and of the necessary theoretical dualism it implied between the real and money economies.

None of the conventional approaches to historiography and the development of knowledge account for, or comment on, the relative centrality of some players and obscurity of others. Why, for example, is Ricardo, who was by no means the most adept theorist, so important and Thornton, Blake, Cobbett, Mr. ****, *et al.* relatively forgotten men? Cultural theory alerts us to the importance of participation in the key social struggles played out behind the theoretical ones, to the importance of the City plays to the bullion controversy, and the inter-relation of social relations and cultural bias or styles of thought. The centrality of players in these struggles provides them with the cultural bias, built of participation and justification, that assures centrality in the theoretical struggles that both overarch and underpin the social ones. The construction of the three-dimensional model alerts us to the importance of actors who bridge the space between the extreme styles of thought and ways of life. In terms of the socio-cultural space it emerges again and again that the central are central. Hence, cultural theory provides an insight into the mystery of why some people emerge as key figures while others do not. It offers a cultural explanation, largely independent of skills and knowledge.

Dispositional approaches based on the attribution of interests, in any direct sense, also fail us in the case of the bullion controversy. Cultural theory alerts us to the weaknesses of both the for and against cases in the controversy surrounding the Silberling thesis. It reveals that it is not a question of any one or more set(s) of interests, but rather a question of the whole justificatory schema or sense-making framework of the way of life of the key participants. We find that what motivated Ricardo was not his pecuniary interest à la Silberling, but his centrality and overwhelming commitment to the social group centerd on the London Stock Exchange.

It was noted throughout that persistent variations in meaning prevented agreement as to the facts. Mannheim's sociology of knowledge suggests that the basic key to the identification of styles of thought is the isolation of systematic variation of meaning between groups. Having traced the outline of the bullion controversy, it is clear that the key definition, that of *money,* varied systematically. This is indicative of the correspondence of bullionist and anti-bullionist ideas with conflicting styles of thought. Bullionists defined money narrowly, focusing on currency, while anti-bullionists defined it broadly: currency, bills of exchange, etc. The former reflects a cultural bias towards simplicity, and the latter a bias towards complexity.

It is obvious that bullionism was no more or less *true* and/or *valid* in 1819 than it had been in 1810. Bullionism was simply the theory which best fit and complemented the style of thought emergent from the lived-experience of the ascendant class and the ascendant group(s) within that class. As Mannheim remarked:

> A theory . . . is wrong if in a given practical situation it uses concepts and categories which, if taken seriously, would prevent man from adjusting himself at that historical stage.[10]

And, as this study reveals, a theory is right when it does allow man to adjust, and, we suggest, *because* it does.

It is a core precept of this study that knowledge and context, cognition and action cannot be separated in a dynamical model. Indeed they must be seen in a mutually (re)constitutive inter-relation. Therefore, knowledge and context are treated together and inter-relatedly. This should not, however, disguise the fact that we have treated specific elements of knowledge. Some specific examples may help to clarify and establish this point.

Amongst the theoretical knowledge elements in the development of monetary theory that have been explicitly discussed are the quantity theory of money, the Real Bills Doctrine, the definition of the theoretical term *money,* the development of the static versus dynamic approach, the development of the abstract deductive versus institutional approach, and the relative analytical separation of real (commodity) and money economies. What our treatment of these suggests is that such elements arise from sets of circumstances and choices, the determinants of which are to be found in the social context of their arising, and *not* from the purely internal factors in knowledge.

It was, for example, revealed that Ricardo adhered to a simplistic quantity theory of money that did *not* embrace the available velocity of circulation argument, which it must be noted would have been entirely compatible with it. It was suggested that this choice (conscious or otherwise) arose in the context of Henry Thornton's apparent support for restriction, that Ricardo simply rejected Thornton's analysis *in toto.* It is a conventional wisdom that Ricardo's move in this regard hindered the later development of monetary theory, such that its impact is both clear and well acknowledged. The same could be said of the case of Ricardo's victory in the face of Baring's dynamic institutional approach circa 1819.

One further example of the way in which the socio-culturally selective choices have affected the development of the analytical detail of monetary theory is that of the analytical separation of the real and money economies. It has been revealed that Ricardo's narrow definition of money, and, it is suggested, his desire to blame the Bank of England, led to his rigid analytical separation of the real and money economies. This separation came to profoundly influence the theoretical development of political economy, because the rigidly demarcated money economy implied—required for coherence—a rigidly demarcated real (commodity) economy. This, together with the influence that Ricardo's ideas carried after the institutionalization of his plan in Peel's Bill, led to the development and institutionalization of the "Corn Model" as the very basis of political economy for at least the next 50 years.

Such examples of the treatment given to contextually sited knowledge in this study could easily be multiplied. The point is, however, simply to unequivocally demonstrate that we have explored the production and development of economic knowledge—albeit unusually contextually.

A further source of obfuscation of the treatment of knowledge may have arisen because the bullion controversy often involved the parallel treatment of economic knowledge and economic policy. It must be noted, however, that both *principles and policy* have been explored. It was noted that, while bullionist principles and policy were *both* rejected in 1811 and *both* accepted in 1819, a separation of them was possible. It would, for example, have been quite coherent for the pro-restriction lobby to have argued in 1811 that it agreed with bullionism in principle, but realized that specie payments could not be made, because of war demands, and/or while exchanges were unfavorable, and/or while Bank reserves were insufficient; and one could not, therefore, enact bullionist policy. The lobby did *not* do this. Rather it produced a series of resolutions concerning the Real Bills Doctrine, together with an alternative definition of money. The lobby was, in short, shown to adhere to an alternative theory, built of a different style of thought, that was independent and complete in itself.

This example also reveals a key hypothesis, namely, that both the development of knowledge *and* its practical application must be treated in such a way as to reveal the extent of its historical contextuality. A major means towards this end is revealed to be the identification of broad styles of thought, by documentary methods, wherein the specific analytical and policy pronouncements can be shown to be the outcome of, and an ongoing transformational force acting upon, the structure of social relations—relations of power—in societies.

In summary, cultural theory and the development of socio-cultural analysis has given us an approach to the study of the development and diffusion of knowledge that is proven to be superior to conventional historiographic approaches. In all the cases noted in this concluding section, and in all those noted in passing during our review of the bullion controversy, cultural theory has passed the test of practice.

This case study drew out the various macro and micro level elements that entered, both into the success, and the failure of bullionism as theory and as policy. In a broad way this was intended to demonstrate the necessity of a wider than scientific community contextual siting, and the importance of developing an appropriate sociology of knowledge. What it has also shown is that rational reconstruction is a partial and incomplete story, and that it is wrong on specifics of the, so-called, logic of scientific discovery and theory choice.

In this particular case the key to unearthing the real underlying bullion controversy story was an analysis of group dimension movement. Pursuing the lead which the socio-cultural schema suggested, in terms of factional conflict, the bullion controversy was set against an increasing social experience of factionalism. How the daily life of David Ricardo put him at the forefront of this movement, and thereby explained why he was, and is, seen as the leader of the new style of analysis, as well as of its substantive content, was also noted. Ricardo was the most bullionist in both stylistic and substantive terms, because he amongst his peers was the most factionally committed to the individualist way of life.

This focus on the group dimension is not necessarily going to be the case for other enquiries. In other instances grid dimension changes might be the most important, such that the focus would be on changes in the regulative structure(s), and on the hierarchical re-organization of social life. This would lead directly to analysis of, for example, the implications of "deregulated" economic environments for economic behavior, or to an analysis of the effects of the organization flattening effects of communications and information technologies, and to other such important research questions.

There have emerged areas in which further work is required in the operationalization and refinement of the three-dimensional model suggested. Further work is required in operationalization of the dimensions—grid, group and power. As was noted in Chapter 2, the grid and group dimensions have already been developed to a stage at which formal operationalization is possible.[11] Much more difficult work remains in regard to the power dimension. There also emerges a need to further develop the dynamics of the model. As it stands here we have undertaken a process involving initial placement followed by tracing relative movement in grid-group-power space. There is a possible problem, however, regarding time scale. Effectively what happens is that the bounds that initially define grid-group-power space themselves change with the passage of time and the impact of historical change. At certain points the defined social space needs to be moved, by zooming and scaling and/or by displacement, if it is to continue to accurately caputure the contextual confines of events.

In general, however, we satisfy ourselves that cultural theory has proven superior to conventional historiography in the case of the bullion controversy. It offers a practicable approach to the study of knowledge development and diffusion that permits the endogenous treatment of gradual and revolutionary change, and uncovers the causal link between underlying social relations and the styles or modes of thought that shape theory and act as the justificatory support of the ways of life that are those given social relations. Such an approach promises to open up the study of knowledge. It can contribute to the analysis of theory choice, to the analysis of theoretical schemas and of the content and analytical structure and detail of theories. In a parallel manner cultural theory proves a rewarding approach to the study of the policy process. It exposes the mechanisms and synergies underlying the development of policy from expert knowledge(s), the generation of political agenda, and the nodes of decision making and choice. It opens for study the inter-relations of policy process and political cultures.

We began by asking:

> How do people know what they know? Why, as all have access to the same nature, physical and human, don't they come to the same conclusions? Or, if each individual is different, why don't they come to wholly different conclusions?[12]

By developing a cultural theory approach we are able to reveal that it is not a matter of choice between the conventional alternatives of realism (empirical realism) and relativism, but a recognition of a five-way pluralism, a simultaneous multi-culturalism. These five ways of life each consist of a matched set of social relations and cultural bias—shared values and beliefs. They are sustained by the generation of preferences from those social relations that reinforce and reproduce those relations. In making sense of what one does, and doing what makes sense, one necessarily brings into line the structure of social relations in which one lives and the cultural bias or filter through which one sees. Hence, as Durkheim put it, the structure of symbolism parallels the structure of social life.

Our perceptions of nature and of human nature, of risk and reward, of rational and irrational, and of right and wrong are the products of the way we live and the way we relate to others. On the basis of analyzing actors' positions on the three key dimensions of sociality—grid, group and power—we can assign social actors to one of the five regions that corresponds to the five sustainable ways of life, and correlate their social relations with their characteristic cultural bias. This correlation forms the basis for an examination of *both* the social relations and the cultural bias. We can, that is, explore the sense of the social relations, and the relations of the various ways of making sense. Thus we can explore both the schemas of thought and the schemas of action; theory and policy.

NOTES

1. R. Bhaskar, 1978a, *A Realist Theory of Science* (Sussex: Harvester), pp. 24–25; S. Clegg, 1983, "Phenomenology and Formal Organizations: A Realist Critique," *Research in Sociology of Organizations,* Vol. 2, p. 125f.; etc.
2. F. H. Knight, 1956, *On the History and Method of Economics* (Chicago: Chicago University Press).
3. P. Sraffa and M. H. Dobb, 1951–55, Eds., *The Works and Correspondence of David Ricardo* (Cambridge: Cambridge University Press), Vol. VI.
4. Forms of logical positivism and logical empiricism have been developed as confirmationist, verificationist and falisificationist models of the development of knowledge.
5. Popperian hypotheses, or Lakatosian scientific research programmes.
6. F. W. Fetter, 1965a, *The Development of British Monetary Orthodoxy, 1797–1875* (Cambridge, Mass: Harvard University Press).
7. Following Lakatos.
8. *Quarterly Review,* 1811, Vol. 5, No. 9, Art. xi, p. 246.
9. See F. W. Fetter, 1965a, op. cit., p. 91, and B. Gordon, 1976, *Political Economy in Parliament 1819–1823* (London: MacMillan), p. 26.

10. K. Mannheim, quoted P. Hamilton, 1974, *Knowledge and Social Structure; An introduction to the Classical Argument in the Sociology of Knowledge* (London: Routledge and Kegan Paul), p. 103.

11. J. L. Gross, and S. Rayner, 1985, *Measuring Culture; A Paradigm for the Analysis of Social Organization* (New York: Columbia University Press).

12. M. Thompson, R. Ellis and A. Wildavsky, 1990, *Cultural Theory* (Boulder CO: Westview Press), p. 129.

Bibliography

Abir-Am, P. 1984. "Beyond Deterministic Sociology and Apologetic History." *Social Studies of Science,* 14(1), pp. 252–63.

Acworth, A. W. 1925. *Financial Reconstruction in England, 1815–1822.* London: King and Sons.

Apter, D. E. 1964. (Ed.). *Ideology and Discontent.* New York: Free Press of Glencoe.

Aron, A. 1989. "The Early Tooke and Ricardo: A Political Alliance and First Signs of Theoretical Disagreements." *History of Political Economy,* 21(1), pp. 1–14.

Ashton, T. S., and Sayers, R. S. 1953. (Eds.), *Papers in English Monetary History.* Oxford: Clarendon Press.

Asromourgos, T. 1988. "The Life of William Petty in Relation to His Economics: A Tercentenary Interpretation." *History of Political Economy,* 20(3), pp. 337–356.

Bain, A. 1966. *James Mill: A Biography.* New York: A. M. Kelly Reprints.

Barnes, B. 1974. *Scientific Knowledge and Sociological Theory.* London: Routledge and Kegan Paul.

______. 1981. "On the 'Hows' and 'Whys' of Cultural Change." *Social Studies of Science,* 11(4), pp. 481–93.

______. 1985. "Ethnomethodology as Science." *Social Studies of Science,* 15, pp. 751–62.

Barnes, D. G. 1961. *A History of the English Corn Laws, 1660–1846.* New York: A. M. Kelly.

Beaugrand, P. 1982. "Henry Thornton: A Mise au Point." *History of Political Economy,* 14(1), pp. 101–111.

Ben David, J. 1975. "Innovations and Their Recognition in Social Science." *History of Political Economy,* 7(4), pp. 434–54.

Berg, Maxine. 1980. *The Machinery Question and the Making of Political Economy, 1815–1848.* Cambridge: Cambridge University Press.

Berger, P. L., and Luckmann, T. 1966. *The Social Construction of Reality: A Treatise in the Sociology of Knowledge.* New York: Doubleday.

Berman, M. 1978. *Social Change and Scientific Organisation: The Royal Institution 1799–1844.* London: Heinemann.

Bhaskar, R. 1975. "Feyerabend and Bachelard: Two Philosophies of Science." *New Left Review,* 94, pp. 31–55.

______. 1978a. *A Realist Theory of Science.* Sussex: Harvester Press.

______. 1978b. *The Logic of Scientific Discovery.* Sussex: Harvester Press.

______. 1979. *The Possibility of Naturalism: A Philosophical Critique of the Contemporary Human Sciences.* Sussex: Harvester Press.

Blake, W. 1810. *Observations on the Principles Which Regulate the Course of Exchanges: and on the Present Depreciated State of the Currency.* London. In J. R. McCulloch (Ed.), 1966, pp. 477–563.

Blaug, M. 1958. *Ricardian Economics, A Historical Study.* New Haven, Conn: Yale University Press.

______. 1975. "Kuhn Versus Lakatos: Or Paradigms Versus Research Programmes in the History of Economics." *History of Political Economy,* 7(4), pp. 399–433.

______. 1976. "Kuhn Versus Lakatos: Or Paradigms Versus Research Programmes in the History of Economics." In S. J. Latsis (Ed.), 1976, pp. 149–80.

______. 1978. *Economic Theory in Retrospect.* Cambridge: Cambridge University Press.

______. 1980. *The Methodology of Economics: Or How Economists Explain.* Cambridge: Cambridge University Press.

Bloomfield, A. I. 1978. "The Impact of Growth and Technology on Trade in Nineteenth-Century British Thought." *History of Political Economy,* 10(4), pp. 608–35.

______. 1981. "British Thought on the influence of Foreign Trade and Investment on Growth 1800–1880." *History of Political Economy,* 13(1), pp. 95–120.

Bloor, Celia, and D. 1982. "Twenty Industrial Scientists: A Preliminary Exercise." In Douglas, Mary, Ed., 1982, pp. 83–102.

Bloor, D. 1976. *Knowledge and Social Imagery.* London: Routledge and Kegan Paul.

______. 1983. *Wittgenstein. A Social Theory of Knowledge.* London: MacMillan.

Blume, S. S. 1977. (Ed.). *Perspectives in the Sociology of Science.* Chichester: John Wiley and Sons.

Bonar, J. 1885. *Malthus and His Work.* London: MacMillan.

Bosanquet, C. 1810. *Practical Observations on the Bullion Report.* London: J. M. Richardson.

Bottomore, T. B. 1975. *Sociology as Social Criticism.* London: George Allen and Unwin.

Boulding, K. 1956. *The Image.* Ann Arbor: University of Michigan Press.

Bowden, G. 1985. "The Social Construction of Oil Reserve Estimates." *Social Studies of Science.* 15, pp. 207–40.

Bowering, J. 1843. *The Works of Jeremy Bentham.* 11 Volumes, Edinburgh: William Tait.

Bowley, Marian. 1967. *Nassau Senior and Classical Economics.* London: George Allen and Unwin.

Brandis, R. 1967. "On the Noxious Influence of Authority." *Quarterly Review of Economics and Business,* 7, pp. 37–44.

Brannigan, A. 1981. *The Social Basis of Scientific Discoveries.* Boston: Cambridge University Press.

Brebner, J. B. 1954. "Laissez-Faire and State Intervention in Nineteenth-Century Britain." In E. M. Carus-Wilson, Ed., 1954, pp. 252–62.

Brenner, M., et al. 1978. Eds., *The Social Contexts of Method.* London: Croom Helm.

Briggs, A. 1948. "Thomas Attwood and the Economic Background to the Birmingham Political Union." *Cambridge Historical Journal,* 9(2), pp. 190–216.

______. 1959. *The Age of Improvement, 1784–1867.* London: Longman.

Bronfenbrenner, M. 1971. "The Structure of Revolutions in Economic Thought." *History of Political Economy,* 3(1), pp. 136–51.

Bukharin, N. I., et al. 1971. *Science at the Cross Roads: Papers Presented to the International Congress of the History of Science and Technology.* London: Frank Cass.

Bullion Committee. 1810. "Report, from the Select Committee of the House of Commons, on the High Price of Gold Bullion." In J. R. McCulloch, Ed., 1966, pp. 403–74.

Burtt, E. J. 1972. *Social Perspectives in the History of Economic Theory.* New York: St Martins Press.

Butlin, S. J. 1951. (Ed.). *Report on the High Price of Bullion with Minutes of Evidence.* Sydney: The University of Sydney Press.

Caldwell, B. J. 1982. *Beyond Positivism: Economic Methodology in the Twentieth Century.* London: George Allen and Unwin.

______. 1984. (Ed.). *Appraisal and Criticism in Economics: A Book of Readings.* Boston: Allen and Unwin.

Callan, M., and Law J. 1982. "On Interests and Their Transformation: Enrolment and Counter-Enrolment." *Social Studies of Science,* 12(4), pp. 615–625.

Campbell, B. L. 1985. "Uncertainty as Symbolic Action in Disputes Among Experts." *Social Studies of Science,* 15, pp. 429–53.

Cannan, E. 1917. *A History of the Theories of Production and Distribution in English Political Economy from 1776 to 1848.* London: Staples Press.

______. 1925. *The Paper Pound of 1797–1821: A Reprint of the Bullion Report with Introduction.* London: P.S. King and Son.

Carr, E. H. 1964. *What is History?* Harmondsworth: Penguin.

Carus-Wilson, E. M. 1954. (Ed.). *Essays in Economic History: Reprints Edited for the Economic History Society.* (3 vols) London: Edward Arnold.

Checkland, S. B. 1948. "The Birmingham Economists." *Economic History Review,* 1(1), pp. 1–19.

Child, A. 1944. "The Problem of Imputation Resolved." In J. E. Curtis, and J. W. Petras, (Eds.), 1970, pp. 668–85.

Clapham, J. 1977. *The Bank of England: A History.* 2 volumes, Cambridge: Cambridge University Press.

Clegg, S. 1978. "Method and Sociological Discourse." In M. Brenner, et al. (Eds.), 1978, pp. 67–90.

______. 1979. *The Theory of Power and Organization.* London: Routledge and Kegan Paul.

______. 1983. "Phenomenology and Formal Organizations: A Realist Critique." *Research in Social Organizations,* 2, pp. 109–52.

______. 1989. *Frameworks of Power.* London: Sage.

Clive, J. 1957. *Scotch Reviewers: The Edinburgh Review 1802–1815.* London: Faber and Faber.

Coats, A. W. 1965. "The Role of Authority in the Development of British Economics." *Journal of Law and Economics,* 7, pp. 85–106.

______. 1969. "Is there a 'Structure of Scientific Revolutions' in Economics?" *Kyklos,* 22, pp. 289–96.

______. (Ed.). 1971. *The Classical Economists and Economic Policy.* London: Methuen and Company.

______. (Ed.). 1983. *Methodological Controversy in Economics: Essays in Honour of T.W. Hutchisn.* Greenwich: JAI Press.

Cobbett, W. 1810–11. "Paper Against Gold: Being an Examination of the Report of the Bullion Committee." *Political Register,* London. (Volumes 18 through 20).

Cole, G. D. H. 1971. *The Life of William Cobbett.* Westport, Conn.: Greenwood Press.

Cole, K., Cameron, J., and Edwards, C. 1983. *Why Economists Disagree: The Political Economy of Economics.* London: Longman.

Collins, H. (Ed.). 1969. *Thomas Paine: The Rights of Man.* Harmondsworth: Penguin.

Collison Black, R. D. 1971. *Readings in the Development of Economic Analysis, 1776–1848.* Newton Abbot: David and Charles.

Collison Black, R. D., Coats A. W., and Goodwin C. D. W. (Eds.). 1973. *The Marginal Revolution in Economics: Interpretation and Evaluation.* Durham: Duke University Press.

Cook, C., and Stevenson, J. 1980. *British Historical Facts 1760–1830.* London: MacMillan Press.

Cookson, J. E. 1975. *Lord Liverpool's Administration: The Crucial Years 1815–22.* Hamdon, Conn.: Archon Books.

______. 1982. *The Friends of Peace: Anti-War Liberalism in England, 1793–1815.* Cambridge: Cambridge University Press.

Cope, S. R. 1942. "The Goldsmids and the Development of the London Money Market during the Napoleonic Wars." *Economica,* 9, 1942 (May), pp. 180–206.

Crafts, N. F. R. 1985. *British Economic Growth During the Industrial Revolution.* Oxford: Clarendon Press.

Cullen, M. J. 1975. *The Statistical Movement in early Victorian Britain: the Foundations of Empirical Social Research.* New York: Harvester Press.

Cunningham-Wood, J. (Ed.). 1985. *David Ricardo: Critical Assessments.* 4 Volumes, London and Sydney: Croom Helm.

Currie, I. D. 1970. "The Sapir-Whorf Hypothesis." In Curtis, J. E., and Petras, J. W. (Eds.), 1970, pp. 403–21.

Curtis, J. E., and Petras J. W. (Eds.). 1970. *The Sociology of Knowledge: A Reader.* New York: Praeger.

David, P. A. 1975. *Technical Choice, Innovation and Economic Growth: Essays on American and British Experiences in the Nineteenth-Century.* New York: Cambridge University Press.

De Gré, G. 1941. "The Sociology of Knowledge and the Problem of Truth." In Curtis, J. E., and Petras, J. W. (Eds.), 1970, pp. 661–67.

______. 1955. *Science as a Social Institution.* New York: Random House.

Deane, Phyllis. 1978. *The Evolution of Economic Ideas.* Cambridge: Cambridge University Press.

Deane, Phyllis, and Cole W.A. 1980. *British Economic Growth 1688–1959: Trends and Structure.* Cambridge: Cambridge University Press.

Dicey, A. V. 1981. *Lectures on the Relations Between Law and Public Opinion in England During the Nineteenth-Century.* New Bruswick, N.J.: Transaction Books.

Don Vann, J., and Van Arsdel, Rosemary T. 1978. *Victorian Periodicals: A Guide to Research.* New York: Modern Language Assoc.

Doty, C. S. (Ed.). 1973. *Western Civilisation: Recent Interpretations II—From 1715 to the Present.* New York: Thomas Y. Crowell and Company.

Douglas, Mary. 1966. *Purity and Danger: An Analysis of the Concepts of Pollution and Taboo.* London: Routledge and Kegan Paul.

______. 1970. *Natural Symbols: Explorations in Cosmology.* New York: Pantheon Books.

______. 1975. *Implicit Meanings: Essays in Anthropology.* London and Boston: Routledge and Kegan Paul.

______. 1978. "Cultural Bias." London: *Royal Anthropological Institute,* Occasional Paper No. 35.

______. 1982. *In the Active Voice.* London: Routledge and Kegan Paul.

______. (Ed.). 1982. *Essays in the Sociology of Perception.* London: Routledge and Kegan Paul.

Douglas, Mary, and Wildavsky, A. 1982. *Risk and Culture.* London: University of California Press.

Dow, L. A. 1977. "Malthus on Sticky Wages: The Upper Turning Point and General Glut." *History of Political Economy,* 9(3), pp. 303–21.

Dunford, R. W. 1985. "The Problem of Relevant Collectivities." *Social Studies of Science,* 15, pp. 455–74.

Durkheim, E. 1965. *The Elementary Forms of Religious Life: Study in Religious Sociology.* translated by Joseph Ward Swain, New York: Free Press.

Durkheim, E., and Mauss, M. 1963. *Primitive Classification.* translated by R. Needham, Chicago: University of Chicago Press.

Eagly, R. V. (Ed.). 1968. *Events, Ideology and Economic Theory: The Determinants of Progress in the Development of Economic Analysis.* Detroit: Wayne State University Press.

Ekelund, R. B., and Price E. O. 1979. "Sir Edwin Chadwick on Competition and the Social Control of Industry: Railroads." *History of Political Economy,* 11(2), pp. 213–39.

Elias, N., Martins, H., and Whitley, R. (Eds.). 1982. *Scientific Establishments and Hierarchies.* Sociology of the Sciences Year Book, 1982. Dordrecht: Reidel.

Elkana, Y. 1978. "Two Tier Thinking: Philosophical Realism and Historical Relativism." *Social Studies of Science,* 8(3), pp. 309–26.

Elster, J. 1983. *Explaining Technical Change.* Cambridge: Cambridge University Press.

Fay, C. R. 1951. *Huskisson and His Age.* London: Longmans, Green and Company.

Ferguson, C. E. 1973. "The Specialisation Gap: Barton, Ricardo and Hollander." *History of Political Economy,* 5(1), pp. 1–13.

Fetter, F. W. 1942. "The Bullion Report Reexamined." *Quarterly Journal of Economics,* 56, pp. 655–665.

______. 1949. "The Life and Writings of John Wheatley." *Journal of Political Economy,* 50, pp. 357–76.

______. 1953. "The Authorship of Economic Articles in the Edinburgh Review 1802–47." *Journal of Political Economy,* 61, pp. 232–59.

______. 1958. "The Economic Articles in the Quarterly Review and Their Authors 1809–52." (2 Parts). *Journal of Political Economy,* 66, pp. 47–64, and pp. 154–70.

______. 1959. "The Politics of the Bullion Report." *Economica,* XXVI, pp. 99–120.

______. 1960. "The Economic Articles in Blackwood's Edinburgh Magazine and Their Authors 1817–53." (2 Parts) *Scottish Journal of Political Economy,* 7, pp. 85–107, and pp. 213–31.

______. 1962. "Economic Articles in the Westminster Review and Their Authors 1824–51." *Journal of Political Economy,* 76, pp. 570–96.

______. 1965a. *The Development of British Monetary Orthodoxy 1797–1875.* Cambridge, Mass: Harvard University Press.

______. 1965b. "Economic Controversy in the British Reviews 1802–1850." *Economica,* 32, pp. 424–37.

______. 1980. *The Economists in Parliament, 1780–1868.* Durham: Duke University Press.

______. (Ed.). 1955. *The Irish Pound 1797–1826: A Reprint of the Report of the Committee of 1804 of the British House of Commons on the condition of the Irish Currency.* Evanston, Il.: Northwestern University Press.

______. (Ed.). 1957. *The Economic Writings of Francis Horner in the Edinburgh Review 1802–06.* London: London University Press.

______. (Ed.). 1964. *Selected Economic Writings of Thomas Attwood.* London: London University Press.

Feyerabend, P. 1978. *Against Method: Outline of an Anarchist Theory of Knowledge.* London: Verso Press.

Flew, A. (Ed.). 1970. *Malthus' Essay on Population.* Harmondsworth: Penguin.

Fontana, B. 1985. *Rethinking the Politics of Commercial Society: The 'Edinburgh Review' 1802–1832.* Cambridge: Cambridge University Press.

Forbes, D. 1954. "'Scientific' Whiggism: Adam Smith and John Millar." *Cambridge Journal,* 7(11).

Foucault, M. 1970. *The Order of Things: An Archaeology of the Human Sciences History.* London: Edward Arnold.

______. 1972. *The Archeaology of Knowledge.* London: Tavistock.

______. 1979. *Discipline and Punish: The Birth of the Prison.* Harmondsworth: Penguin.

Fraser, L. M. 1947. *Economic Thought and Language. A critique of some Fundamental Economic Concepts.* London: Adam and Charles Black.

Freudenthal, G. 1984. "The Role of Shared Knowledge: The Failure of the Constructivist Programme in the Sociology of Science." *Social Studies of Science,* 14, pp. 285–95.

Friedman, M. 1970. "The Counter Revolution in Monetary Thought." *Institute of Economic Affairs,* Occasional Paper No. 33.

Furner, Mary. 1975. *Advocacy and Objectivity.* Kentucky: Kentucky University Press.

Gash, N. 1981. *Aristocracy and People: Britain 1815–1865.* London: Edward Arnold.

Gaskell, G., and Hampton, J. 1982. "A Note on Styles in Accounting." In Douglas, Mary, (Ed.), 1982, pp. 102–112.

Geertz, C. 1964. "Ideology as a Cultural System." In Apter, D. E., (Ed.), 1964, pp. 47–76.

Gellner, E. 1965. *Thought and Change.* Chicago: University of Chicago Press.

Gide, C., and Rist, C. 1948 (1915). *A History of Economic Doctrines.* London: George G. Harrap and Co.

Gilbert, G. 1980. "Economic Growth and the Poor in Malthus' Essay on Population." *History of Political Economy,* 12(1), pp. 83–96.

Goodwin, C. D. 1980. "Towards a Theory of the History of Economics." *History of Political Economy,* 12(4), pp. 610–19.

Goodwin, C. D. W. 1973. "Marginalism Moves the World." In Collison Black, R. D., Coats, A. W., and Goodwin C. D. W. Eds., 1973, pp. 285–304.

Gordon, B. 1969. "Criticism of Ricardian views on Value and Distribution in the British Periodicals, 1820–1850." *History of Political Economy,* 1(2), pp. 370–87.

______. 1976. *Political Economy in Parliament 1819–1823.* London: MacMillan.

______. 1987. "Alexander Baring versus David Ricardo: Economic Policy and Parliament after 1815." *Occasional Paper 132,* Newcastle: The University of Newcastle.

Gordon, D. F. 1965. "The Role of the History of Economic Thought in the Understanding of modern Economic Theory." *American Economic Review,* 55, pp. 119–27.

Gould, J. M., and Kelley, A. M. (Eds.). 1949. *Lecture Notes on Types of Economic Theory: as delivered by W.C.Mitchell.* (2 Vols) New York: A. M. Kelley.

Gourvitch, A. 1966. *Survey of Economic Theory on Technological Change and Development.* New York: A. M. Kelly.

Grampp, W. D. 1973. "Classical Economics and its Moral Critics." *History of Political Economy,* 5(2), pp. 359–74.

______. 1982. "Economic Opinion When Britain Turned to Free Trade." *History of Political Economy,* 14(4), pp. 496–520.

Gross, J. L., and Rayner, S. 1985. *Measuring Culture: A Paradigm for the Analysis of Social Organization.* New York: Columbia University Press.

Grubel, H. G. 1961. "Ricardo and Thornton on the Transfer Mechanism." *Quarterly Journal of Economics,* 75, pp. 292–301.

Gudeman, S. 1986. *Economics as Culture: Models and Metaphors of livelihood.* London: Routledge and Kegan Paul.

Hacking, I. 1983. *Representing and Intervening.* Cambridge: Cambridge University Press.

Hahn, F., and Hollis, M. 1979. *Philosophy and Economic Theory.* Oxford: Oxford University Press.

Halevy, E. 1952. *The Growth of Philosophical Radicalism.* London: Faber and Faber.

Hamilton, P. 1974. *Knowledge and Social Structure: An introduction to the classical argument in the Sociology of Knowledge.* London: Routledge and Kegan Paul.

Hamond, J. L., and Barbara. 1978. *The Village Labourer.* London: Longman.

Hampton, J. 1982. "Giving the Grid/Group Dimensions and Operational Definition." In Douglas, Mary, (Ed.), 1982, pp. 64–82.

Hands, D. W. 1979. "The Methodology of Economic Research Programmes." *Philosophy of the Social Sciences,* 9, pp. 293–303.

______. 1984. "The Role of Crucial Counterexamples in the Growth of Economic Knowledge: Two case studies in the recent history of economic thought." *History of Political Economy,* 16(1), pp. 59–67.

Hansard, J. 1953. *Hansard's Catalogue and Breviate of Parliamentary Papers, 1696–1834.* Oxford: Blackwell.

Hanson, N. R. 1969. *Perception and Discovery.* San Francisco: Freeman Cooper.

Harré, R., and Madden, E. H. 1975. *Causal Powers: A theory of natural necessity.* Totowa, N.J.: Rowman and Littlefield.

Harsanyi, J. 1966. "The Dimension and Measurement of Power." In Rothschild, K. W., (Ed.), 1966, pp. 77–96.

Hartung, F. E. 1952. "Problems of the Sociology of Knowledge." *Philosophy of Science,* XIX, pp. 17–32.

Hayek, F. A. Von. 1949. *Individualism and the Economic Order.* London: Routledge and Kegan Paul.

______. (Ed.). 1939. *An Enquiry into the Nature and Effects of the Paper Credit of Great Britain (1802) by Henry Thornton.* London: George Allen and Unwin.

Heertje, A. 1977. *Economics and Technical Change.* New York: John Wiley and Sons.

Heidegger, M. 1956. *Existence and Being.* London: Vision Press.

Henderson, J. P. 1983. "The Oral Tradition in British Economics: Influential Economists in the Political Economy Club of London." *History of Political Economy,* 15(2), pp. 149–80.

Hessen, B. 1931. "The Social and Economic Roots of Newton's *Principia.*" In Bukharin, N. I., et al. 1971, pp. 146–212.

Hicks, J. 1969. *A Theory of Economic History.* Oxford: Oxford University Press.

______. 1976. "'Revolutions' in Economics." In Latsis, S. J. Ed. 1976, pp. 207–218.

Hilton, B. 1977. *Corn, Cash and Commerce: The Economic Policies of the Tory Governments 1815–1830.* Oxford: Oxford University Press.

Hindess, B. 1977. *Philosophy and Methodology in the Social Sciences.* Sussex: Harvester.

______. 1986. "Actors and Social Relations." In Wardell, M., and Turner, S. (Eds.), 1986, pp. 113–126.

Hindess, B., and Hirst, P. Q. 1975. *Precapitalist Modes of Production.* London and Boston: Routledge and Kegan Paul.

______. 1977. *Modes of Production and Social Formation: an Auto-critique of Precapitalist Modes of Production.* London: MacMillan.

Hirshleifer, J. 1985. "The Expanding Domain of Economics." *American Economic Review,* 75(6), pp. 53–68.

Hobsbaum, E. J., and Rudé, G. 1973. *Captain Swing.* Harmondsworth: Penguin University Books.

Hollander, S. 1971. "The Development of Ricardo's Position on Machinery." *History of Political Economy,* 3(1), pp. 105–35.

______. 1977. "Ricardo and the Corn Laws: A Revision." *History of Political Economy,* 9(1), pp. 1–47.

______. 1979. *The Economics of David Ricardo.* Toronto: The University Press.

Hollis, M., and Lukes, S. (Eds.). 1982. *Rationality and Relativism.* London: Blackwell.

Hont, I., and Ignatieff, M. (Eds.). 1985. *Wealth and Virtue: The shaping of political economy in the Scottish Enlightenment.* Cambridge: Cambridge University Press.

Horowitz, D. 1974. "Historians and Economists: Perspectives on the Development of American Economic Thought." *History of Political Economy,* 6(4), pp. 454–62.

Horsefield, J. K. 1949. "The Bankers and the Bullionists in 1819." *Journal of Political Economy,* 57, pp. 442–48.

Hoselitz, B. F. 1965. *Theories of Economic Growth.* New York: Free Press.

Houghton, J. W. 1988. "Knowledge, Method and Bullionism: The Grid-Group-Power of Economic Knowledge," unpublished doctoral thesis, Brisbane, Australia: University of Queensland.

Houghton, W. E. (Ed.). 1979. *The Wellesley Index to Victorian Periodicals.* 93 Volumes, Toronto: The University Press.

Hume, D. 1777. *Essays and Treatise on Several Subjects.* In Selby-Bigge, L. A. (Ed.), 1975, Oxford: Clarendon Press.

Hunt, E. K. 1977. "Value Theory in the Writings of the Classical Economists, Thomas Hodgskin and Karl Marx." *History of Political Economy,* 9(3), pp. 322–45.

______. 1979. "Utilitarianism and the Labour Theory of Value: A Critique of the Ideas of William Thompson." *History of Political Economy,* 11(4), pp. 545–71.

Huskisson, W. 1810. "The Question Concerning the Depreciation of Our Currency Stated and Examined." London. In McCulloch, J. R. (Ed.), 1966, pp. 567–668.

Hutchison, T. W. 1953. "James Mill and the Political Education of Ricardo." *Cambridge Journal,* 7(2), pp. 81–100.

______. 1977. *Knowledge and Ignorance in Economics.* Chicago: University of Chicago Press.

______. 1978. *On Revolutions and Progress in Economic Knowledge.* Cambridge: Cambridge University Press.

Irish Currency Committee. 1804. "Report from the Select Committee on the Circulating Paper, the Specie and the Current Coin of Ireland." In Fetter, F. W. (Ed.), 1955.

Johnson, H. G. 1975. *On Economics and Society.* Chicago: University of Chicago Press.

Johnston, R. 1984. "Controlling Technology." *Social Studies of Science,* 14(1), pp. 97–113.

Jones, Greta. 1978. "A Social History of Darwin's Descent of Man." *Economy and Society,* 7(1), pp. 1–23.

Karsten, S. 1983. "Dialectics, Functionalism, and Structuralism in Economic Thought." *American Journal of Economics and Sociology,* 42, pp. 179–92.

Karsten, S. G. 1973. "Dialectics and the Evolution of Economic Thought." *History of Political Economy,* 5(2), pp. 399–419.

Katouzian, H. 1980. *Ideology and Method in Economics.* London: MacMillan.

Kautsky, K. (Ed.). 1954 *Theories of Surplus Value.* London: Lawrence and Wishart, translated: Bonner, G.A., and Burns, E.

Keat, R., and Urry, J. 1975. *Social Theory as Science.* London: Routledge and Kegan Paul.

Kecskemeti, P. (Ed.). 1953. *Essays on Sociology and Social Psychology, By Karl Mannheim.* London: Routledge and Kegan Paul.

King, P. 1983. *The History of Ideas: An introduction to method.* London: Croom Helm.

Kneller, G. F. 1984. *Movements of Thought in Modern Education.* New York: John Wiley and Sons.

Knight, F. H. 1956. *On the History and Method of Economics.* Chicago: University of Chicago Press.

Knorr, K. D., Krohn, R., and Whitley, R. (Eds.). 1980. *The Social Process of Scientific Investigation.* Social Sciences Yearbook, IV, 1980, Netherlands: D.Reidel.

Knorr-Cetina, Karin. D. 1982. "The Constructivist Programme in the Sociology of Science: Retreats and Advances." *Social Studies of Science,* 12(1), pp. 133–36.

Krupp, S. R. (Ed.). 1966. *The Structure of Economic Science: Essays on Methodology.* Englewood Cliffs: Prentice-Hall.

Kuhn, T. S. 1962. *The Structure of Scientific Revolutions.* Chicago: University of Chicago Press.

Lakatos, I, and Musgrave A. 1970. *Criticism and the Growth of Knowledge.* Cambridge: Cambridge University Press.

______ 1976. *Proofs and Refutations.* Cambridge: Cambridge University Press.

Landes, D. S. 1969. *The Unbound Prometheus: Technical Change and Industrial Development in Western Europe from 1750 to the present.* Cambridge: Cambridge University Press.

Langer, G. F. 1987. *The Coming Age of Political Economy, 1815–1825.* Westport CT: Greenwood Press.

Langlois, R. N. (Ed.). 1986. *Economics as a Procress: Essays in the New Institutional Economics.* New York: Cambridge University Press.

Latsis, S. 1972. "Situational Determinism in Economics." *British Journal of Philosophy of Science,* 23, pp. 207–45.

______ (Ed.). 1976. *Method and Appraisal in Economics.* Cambridge: Cambridge University Press.

Laurent, J. 1984. "Science, Society and Politics in Late Nineteenth Century England." *Social Studies of Science,* 14, pp. 585–619.

Law, J. (Ed.). 1986. *Power, Action and Belief: A New Sociology of Knowledge?* London and Boston: Routledge and Kegan Paul.

Layder, D. 1981. *Structure, Interaction and Social Theory.* London: Routledge and Kegan Paul.

Lehmann, W. C. 1960. *John Millar of Glasgow, 1735–1801. His life and Thought and his contributions to Sociological analysis.* Cambridge: Cambridge University Press.

Leigh, I. 1951. *Castlereagh.* London: Collins.

Leijonhufvud, A. 1976. "Schools, 'Revolutions', and Research Programmes in Economic Theory." In Latsis, S. J. (Ed.), 1976, pp. 65–108.

Leiss, W. 1975. "Ideology and Science." *Social Studies of Science,* 5(2), pp. 193–201.

Lemaine, G., et al. (Eds.). 1976. *Perspectives on the Emergence of Scientific Disciplines.* New York: Aldine Press.

Letwin, Shirley. 1965. *The Pursuit of Certainty.* Cambridge: Cambridge University Press.

Levy, D. 1976. "Ricardo and The Iron Law: A Correction to the Record." *History of Political Economy,* 8(2), pp. 235–51.

Liverpool, Earl Of. 1805. "Remarks on Paper Currency: from a Treatise on the Coins of the Realm." London. In McCulloch, J. R. (Ed.), 1966, pp. 350–359.

Lukes, S. 1971. "The Meanings of Individualism." *Journal of the History of Ideas,* 32(1), pp. 45–66.

Lyons, D. 1973. *In the Interest of the Governed.* Oxford: Clarendon Press.

MacKenzie D. 1981. "Interests, Positivism and History." *Social Studies of Science,* 11(4), pp. 498–504.

Machlup, F. 1978. *Methodology of Economics and other Social Sciences.* New York: Academic Press.

Mack, Mary. P. 1962. *Jeremy Bentham: An Odyssey of Ideas 1748–1792.* London: Heinemann.

Malthus, T. R. 1798. "An Essay on the Principle of Population as it affects the Future Improvement of Society with remarks on the Speculations of Mr. Goodwin, M. Condorcet, and other writers." In Flew, A. (Ed.), 1970, pp. 60–217.

_____ 1811. "Pamphlets on the Bullion Question." *Edinburgh Review,* 18(36), pp. 448–470.

Mannheim, K. 1936. *Ideology and Utopia: An Introduction to the Sociology of Knowledge.* London: Routledge and Kegan Paul.

_____ 1953. "Essays on the Sociology of Knowledge." In Kecskemeti, P. (Ed.), 1953, pp. 15–294.

_____ 1970. "The Sociology of Knowledge." Curtis, J. E., and Petras, J. W. (Eds.), 1970, pp. 109–30.

Marcet, Jane. 1827. *Conversations on Political Economy: In Which the Elements of that Science are Familiarly Explained.* London: Longman.

Marchi, N. B. de 1970. John Stuart Mill and the Development of English Economic Thought: A study in the progress of Ricardian orthodoxy. Unpublished doctoral dissertation, Canberra: Australian National University.

_____ 1973. "The Noxious Influence of Authority: A Correction of Jevons." *Journal of Law and Economics,* 16, pp. 179–89.

Martineau, Harriet. 1877–78. *A History of the Thirty Years' Peace.* 4 Volumes, London: George Bell and Sons.

_____ 1893. *Biographical Sketches, 1852–1875.* London: MacMillan.

Mason, W. E. 1957. "Ricardo's Transfer Mechanism Theory." *Quarterly Journal of Economics,* 71, pp. 107–115.

Mathias, P. 1969. *The First Industrial Nation: An Economic History of Britain 1700–1914.* London: Methuen and Company.

_____ P. (Ed.). 1972. *Science and Society 1600–1900.* Cambridge: Cambridge University Press.

Mazlish, B. 1975. *James and John Stuart Mill: Father and Son in the nineteenth century.* New York: Basic Books.

McCloskey, D. 1984. "The Rhetoric of Economics." In Caldwell, B. J. (Ed.), 1984, pp. 320–56.

_____ 1985. *The Rhetoric of Economics.* New York: Harvester Press.

McCulloch, J. R. 1888. *The Works of David Ricardo.* London: Murray.

_____ (Ed.). 1966. *A Select Collection of Scarce and Valuable Tracts and other publications on Paper Currency and Banking.* New York: A.M.Kelley.

McKenzie, R. 1981. "The Necessary Normative Context of Positive Economics." *Journal of Economic Issues,* 15, pp. 703–19.

Meek, R. 1967. *Economics and Ideology, and other essays.* London: Chapman and Hall.

Merton, R. K. 1973. *The Sociology of Science.* Chicago: University of Chicago Press.

Mill, J. 1808. "Commerce Defended." In Winch, D. (Ed.), 1966, pp. 85–159.

_____ 1829. *Analysis of the Phenomena of the Human Mind.* 2 Volumes, London: Longman.

_____ 1966 (1804). "An Essay on the Impolicy of a Bounty on the Exportation of Grain: and on the Principles which ought to Regulate the Commerce of Grain." In Winch D. (Ed.), 1966, pp. 41–84.

Millar, J. 1779. "Origin of the Distinction of Ranks." In Lehmann, W. C. (Ed.), 1960, pp. 165–322.

Miller, D. P. 1984. "Social History of British Science." *Social Studies of Science,* 14(1), pp. 115–35.

Millstone, E. 1978. "A Framework for the Sociology of Knowledge." *Social Studies of Science,* 8(1), pp. 111–125.

Mitchell, W. 1949. "Lecture notes on Types of Economic Theory." In Gould, J. M., and Kelley, A. M. (Eds.), 1949.

Morgan, E. V., and Thomas, W. A. 1962. *The Stock Exchange: Its history and functions.* London: Elek Books.

Moss, D. J. 1981. "Banknotes versus Gold: The Monetary Theory of Thomas Attwood in His Early Writings, 1816–19." *History Of Political Economy,* 13(1), pp. 19–38.

Mulkay, M. 1979a. *Science and the Sociology of Knowledge.* London: George Allen and Unwin.

_____ 1979b. "Knowledge and Utility: Implications for the Sociology of Knowledge." *Social Studies of Science,* 9(1), pp. 63–80.

_____ 1984. "The Scientist Talks Back." *Social Studies of Science,* 14, pp. 265–82.

Mulkay, M., and Gilbert, G. N. 1982. "What is the ultimate Question?, Some Remarks in Defense of the Analysis of Scientific Discourses." *Social Studies of Science,* 12(2), pp. 309–319.

Musson, A. E. (Ed.). 1972. *Science, Technology and Economic Growth in the Eighteenth Century.* London: Methuen.

Myers, M. L. 1972. "Anticipations of Lassez-Faire." *History of Political Economy,* 4(1), pp. 163–75.

Myrdal, G. 1969. *The Political Element in the Development of Economic Theory.* translated by P. Streeton, New York: Clarion Books.

Nabers, L. 1966. "The Positive and Genetic Approaches." In Krupp, S. R. (Ed.), 1966, p. 68ff.

Negishi, T. 1982. "The Labour Theory of Value in the Ricardian Theory of International Trade." *History of Political Economy,* 14(2), pp. 199–210.

O'Brien, D. P. 1970. *J. R. McCulloch: A Study in Classical Economics.* London: George Allen and Unwin.

_____ 1975. *The Classical Economists.* Oxford: Clarendon Press.

_____ 1983. "Theories of the History of Science." In Coats, A. W. (Ed.), 1983, pp. 89–124.

O'Brien, D. P., and Darnell, A. C. 1980. "Torrens, McCulloch and the Digression on Sismondi." *History of Political Economy,* 12(3), pp. 385–95.

O'Driscoll, G. P. 1975. "The Specialisation Gap and the Ricardo Effect." *History of Political Economy,* 7(2), pp. 261–69.

Ostrander, D. 1982. "One- and Two-Dimensional Models of the Distribution of Beliefs." In Douglas, Mary, (Ed.), 1982, pp. 14–30.

Owen, D. E. 1982. "Spectral Evidence: The Witchcraft Cosmology of Salem Village in 1692." In Douglas, Mary, (Ed.), 1982, pp. 275–301.

Paine, T. 1791/2. "The Rights of Man: Being and answer to Mr Burke's attack on the French Revolution." In Collins, H. (Ed.), 1969, pp. 61–131.

Palmer, A. W. 1964. *The Penguin Dictionary of Modern History, 1789–1945.* Harmondsworth: Penguin.

Parry, L. A. 1934. *The History of Torture in England.* London: Sampson Law, Marston and Company.

Peake, C. F. 1978. "Henry Thornton and the Development of Ricardo's Economic Thought." *History of Political Economy,* 10(2), pp. 193–212.

Perlman, M. 1986. "The Bullion Controversy Revisited." *Journal of Political Economy,* 94(4), pp. 745–62.

______ 1989. "Adam Smith and the Paternity of the Real Bills Doctrine." *History of Political Economy,* 21(1), pp. 77–90.

Petrella, F. 1977. "Benthamism and the Demise of the Classical Economic Ordnungspolitik." *History of Political Economy,* 9(2), pp. 215–36.

Pinch, T. J., and Bijker, W. E. 1984. "The Social Construction of Facts and Artefacts." *Social Studies of Science,* 14(3), pp. 399–441.

Pinch, T. J., and Collins, H. M. 1984. "Private Science and Public Knowledge." *Social Studies of Science,* 14, pp. 521–46.

Plamenatz, J. 1949. *The English Utilitarians.* Oxford: Basil Blackwell.

Polanyi, M. 1967. *The Tacit Dimension.* New York: Doubleday.

Political Economy Club. 1882. *Minutes of Proceedings of Meetings.* IV. London: MacMillan.

______ 1921. *Centenary Volume: VI.* London: MacMillan.

Pollard, S. 1968. *The Genesis of Modern Management.* Harmondsworth: Penguin.

Powell, E. T. 1916 (1915). *The Evolution of the Money Market, 1385–1915.* London: Financial News.

Price, L. L. 1931. *A Short History of Political Economy in England.* London: Methuen.

Putnam, H. 1981. *Reason, Truth and History.* Cambridge: Cambridge University Press.

Rae, J. 1834. *Statement of some New Principles on the Subject of Political Economy.* New York: A. M. Kelly. (1964).

______ 1895. *Life of Adam Smith.* London: MacMillan.

Rashid, S. 1981. "Malthus' Principles and British Economic Thought 1820–1835." *History of Political Economy,* 13(1), pp. 55–79.

______ 1983. "Edward Copleston, Robert Peel and Cash Payments." *History of Political Economy,* 15(2), pp. 249–59.

Ravetz, J. R. 1971. *Scientific Knowledge and it Social Problems.* Oxford: Clarendon Press.

Rayner, S. 1989. "Risk, Uncertainty and Social Organization." *Contemporary Sociology,* 18(1), pp. 6–9.

______ 1986. "The Politics of Schism: Routinisation and Social Control in the International Socialists'/Socialist Workers' Party." In Law, J. (Ed.), 1986, pp. 46–67.

Redner, H. 1987. *The Ends of Science: An Essay in Scientific Authority.* Boulder, Colorado: Westview Press.

Reisman, D. A. 1971. "Henry Thornton and Classical Monetary Economics." *Oxford Economic Papers,* 23, pp. 70–89.

Ricardo, D. 1810. *The High Price of Bullion: A Proof of the Depreciation of Banknotes.* In Sraffa, P., and Dobb, M. H. (Eds.), 1951–55, Volume III.

______ 1810. *The High Price of Bullion: A Proof of the Depreciation of Banknotes.* In McCulloch, J. R. (Ed.), 1966, pp. 363–401.

______ 1809. *The High Price of Gold.* In Sraffa, P., and Dobb, M. H. (Eds.), 1951–55, Volume III.

______ 1811. *Reply to Mr. Bosanquet's Practical Observations on the Bullion Report.* In Sraffa, P., and Dobb, M. H. (Eds.), 1951–55, Volume III.

______ 1815. *Proposals for an Economical and Secure Currency: With Observations on the Profits of the Bank of England.* In Sraffa, P., and Dobb, M. H. (Eds.), 1951–55, Volume III.

______ 1819 *On the Principles of Political Economy and Taxation.* London: John Murray.

______ 1821. *On the Principles of Political Economy and Taxation.* In Sraffa, P., and Dobb, M. H. (Eds.), 1951–55, Volume 1.

Robbins, L. 1952. *The Theory of Economic Policy in English Classical Political Economy.* London: MacMillan.

______ 1963. *Politics and Economics: Paper in Political Economy.* London: MacMillan.

Roberts, M. 1935a. "The Fall of the Talents, March 1807." *English Historical Review,* 50, pp. 61–77.

______ 1935b. "The Leadership of the Whig Party in the House of Commons 1807 to 1815." *English Historical Review,* 50, pp. 620–638.

Rogin, L. 1956. *The Meaning and Validity of Economic Theory.* New York: Ayer and Company.

Roll, E. 1954. *A History of Economic Thought.* London: Faber and Faber.

Romano, R. M. 1982. "The Economic Ideas of Charles Babbage." *History of Political Economy,* 14(3), pp. 385–405.

Rose, H and S. 1969. *Science and Society.* Harmondsworth: Penguin.

Rosenberg, N. 1975. "Problems in the Economists' Conceptualisation of Technological Innovation." *History of Political Economy,* 7(4), pp. 456–81.

Roth, M. S. 1981. "Foucault's History of the Present." *History and Theory.* 20.

Rothschild, K. W. (Ed.). 1971. *Power in Economics, Selected Readings.* Harmondsworth: Penguin.

Routh, G. 1975. *The Origin of Economic Ideas.* London: MacMillan.

Rudwick, M. 1982. "Cognitive Styles in Geology." In Douglas, Mary, (Ed.), 1982, pp. 219–42.

Ryan, C. C. 1981. "The Friends of Commerce: Romantic and Marxist Criticisms of Classical Political Economy." *History of Political Economy,* 13(1), pp. 80–94.

Samuels, W. J. 1974. "The History of Economic Thought as Intellectual History." *History of Political Economy,* 6(3), pp. 305–23.

_____ (Ed.). 1980, *The Methodology of Economic Thought.* New Jersey: Transaction Books.

_____ (Ed.). 1983/85. *Research in the History of Economic Thought and Methodology.* 2 Volumes, Greenwich: JAI Press.

Santiago-Valiente, W. 1988. "Historical Background of the Classical Monetary Theory and the 'Real-Bills' Banking Tradition." *History of Political Economy,* 20(1), pp. 43–63.

Sartre, J-P. 1968 (1905). *Search for a Method.* New York: Vintage Books.

Say, J-B. 1964. *A Treatise on Political Economy, or the Production, Distribution and Consumption of Wealth.* translated by C. R. Princep. New York: A. M. Kelley.

Sayers, R. S. 1953. "Ricardo's Views on Monetary Questions." In Coats, A. W. (Ed.), 1971, pp. 33–56.

Sayers, S. 1985. *Reality and Reason: Dialectic and the Theory of Knowledge.* Oxford: Blackwell.

Scheler, M. 1970. "The Sociology of Knowledge: Formal Problems." In Curtis, J. E., and Petras, J. W. (Eds.), 1970, pp. 170–86.

Schofield, R. E. 1967. *The Lunar Society of Birmingham: A social history of provincial science and industry in eighteenth century England.* Oxford: Clarendon Press.

Schumpeter, J. A. 1912. *Economic Doctrine and Method: an historical sketch.* translated by R. Aris, London: George Allen and Unwin.

_____ 1954. *The History of Economic Analysis.* Oxford: Oxford University Press.

Selby-Bigge, L. A. (Ed.). 1975. *Hume's Enquiries Concerning Human Understanding and Concerning the Principles of Morals.* Oxford: Clarendon Press.

Seligman, B. B. 1969. "The Impact of Positivism on Economic Thought." *History of Political Economy,* 1(2), pp. 256–78.

Shackle, G. L. S. 1973. *Epistemics and Economics: A Critique of Economic Doctrines.* Cambridge: Cambridge University Press.

_____ 1979. *Imagination and the Nature of Choice.* Edinburgh: University of Edinburgh Press.

Shapin, S. A. 1972. "The Pottery Philosophical Society: 1819 to 1835." *Social Studies of Science,* 2(3), pp. 311–336.

_____ 1980. "A Course in the Social History of Science." *Social Studies of Science,* 10(2), pp. 231–258.

Shapin, S. A., and Barnes, B. 1977. "Science, Nature and Control: Interpreting Mechanics' Institutes." *Social Studies of Science,* 7(1), pp. 31–74.

Sheridan-Smith, A. M. 1980. *Michel Foucault: The Will to Truth.* London: Tavistock.

Shiner, L. 1982. "Reading Foucault: Anti-Method and the Genealogy of Power-Knowledge." *History and Theory,* 21. p. 382ff.

Shrum, W. 1984. "Scientific Specialities and Technical Systems." *Social Studies of Science,* 14, pp. 63–90.

Silberling, N. J. 1919. "The British Financial Experience, 1790–1830." *The Review of Economic Statistics,* 1(4), October 1919, pp. 287–291.

______ 1924. "The Financial and Monetary Policy of Great Britain during the Napoleonic Wars." 2 Parts, *Quarterly Journal of Economics,* 38, pp. 214–33 and pp. 397–439.

Simonds, A. P. 1978. *Karl Mannheim's Sociology of Knowledge.* Oxford: Clarendon.

Sismondi, J. C. L. Simonde De. 1966. *Political Economy and the Philosophy of Government.* New York: A. M. Kelley.

Skinner, A. S. 1974. *Adam Smith and the Role of the State.* Glasgow: Glasgow University Press.

______ 1979. *A System of Social Science, Papers relating to Adam Smith.* Oxford: Clarendon Press.

Skinner, Q. 1969. "Meaning and Understanding in thee History of Ideas." *History and Theory,* 8, pp. 3–53.

Smart, J. J. C., and Williams, B. 1973. *Utilitarianism: For and Against.* Cambridge: Cambridge University Press.

Smith, A. 1776. *An Inquiry into the Nature and Causes of the Wealth of Nations.* 2 Volumes, London: Strahan and Cadell.

______ 1983. *An Inquiry into the Nature and Causes of the Wealth of Nations.* Harmondsworth: Penguin.

______ 1966. *The Theory of Moral Sentiments or An Essay Towards the Analysis of the Principles by which Men Naturally Judge Concerning the Conduct and Character, First of their Neighbours, and Afterwards of Themselves.* New York: A. M. Kelley.

Sowell, T. 1972. *Say's Law. An Historical Analysis.* Princeton: The University Press.

______ 1974. *Classical Economics Reconsidered.* Princeton: The University Press.

Speier, H. 1938. "The Social Determination of Ideas." *Social Research,* 5, pp. 182–205.

Spengler, J. J. 1968. "Exogenous and Endogenous Influences on the Formation of Post-1870 Economic Thought: A Sociology of Knowledge Approach." In Eagly, R. V. (Ed.), 1968, pp. 159–87.

Sraffa, P., and Dobb, M. H. (Eds.). 1951 to 1955. *The Works and Correspondence of David Ricardo.* 10 volumes, Cambridge: Cambridge University Press.

Stark, W. 1944. *The History of Economics in Its Relation to Social Development.* London: Kegan, Paul, Trench, Trubner and Co.

______ 1958. *The Sociology of Knowledge: An Essay in Aid of a Deeper Understanding of the History of Ideas.* London: Routledge and Kegan Paul.

Steedman, I. et al. 1981. *The Value Controversy.* London: Verso Press.

Stephen, L. 1900. *The English Utilitarians.* 3 Volumes, London: Duckworth.

Stephen, L., and Lee, S. (Eds.). 1917. *The Dictionary of National Biography.* 22 Volumes, Oxford: Oxford University Press.

Stigler, G. J. 1965. *Essays in the History of Economics.* Chicago: University of Chicago Press.

______ 1969. "Does Economics Have a Useful Past?" *History of Political Economy,* 1(2), pp. 217–30.

______ 1973. "The Adoption of Marginal Utility Theory." In Collison Black, R. D., Coats, A. W., and Goodwin, C. D. W. (Eds.), 1973, pp. 305–20.

______ 1982. *The Economist as Preacher, And Other Essays.* Oxford: Blackwell.

Stillinger, J. (Ed.). 1969. *J. S. Mill's Autobiography, And Other Writings.* Boston: Houghton Mifflin.

Sturges, R. P. 1982. "The Career of John Barton: Economist and Statistician." *History of Political Economy,* 14(3), pp. 366–84.

Suppe, F. (Ed.). 1979. *The Structure of Scientific Theories.* Urbana: University of Illinois Press.

Theocharis, R. D. 1961. *Early Developments in Mathematical Economics.* London: MacMillan.

Thom, R. 1975. *Structural Stability and Morphogenesis. An Outline of a General Theory of Models.* translated by R. Thom and D. H. Fowler, Reading, Mass.: W. A. Benjamin.

Thomas, W. 1979. *The Philosophic Radicals: Nine Studies in Theory and Practice 1817–1841.* Oxford: Clarendon Press.

Thomis, M. I. 1972. *The Luddites: Machine-Breaking in Regency England.* New York: Schocken.

Thompson, E. P. 1963. *The Making of the English Working Class.* London: Gollancz.

Thompson, M. 1979. *Rubbish Theory. The creation and Destruction of Value.* Oxford: Oxford University Press.

______ 1982a. "A Three Dimensional Model." In Douglas, Mary, Ed. 1982, pp. 31–62.

______ 1982b. "The Problem of Centre: An Autonomous Cosmology." In Douglas, Mary, (Ed.), 1982, pp. 302–328.

Thompson, N. W. 1984. *The People's Science: The Popular Political Economy of Exploitation and Crisis 1816–34.* Cambridge: Cambridge University Press.

Thornton, H. 1802. *An Enquiry into the Nature and Effects of the Paper Credit of Great Britain.* In McCulloch, J. R., (Ed.), 1966, pp. 140–340.

Thweatt, W. O. 1974. "The Digression on Sismondi: by Torrens or McCulloch." *History of Political Economy,* 6(4), pp. 435–53.

______ 1980. "Torrens, McCulloch and the 'Digression on Sismondi': Whose Digression?" *History of Political Economy,* 12(3), pp. 396–410.

Toulmin, S. 1973. *Human Understanding.* Oxford: Clarendon Press.

Tonnies, F. 1887. *Fundamental Concepts of Sociology.* translated by C. P. Loomis, New York: American Book Company.

Tribe, K. 1978. *Land, Labour and Economic Discourse.* London: Routledge and Kegan Paul.

______ 1981. *Genealogies of Capitalism.* London: MacMillan.

______ 1981. "Ricardian Histories." *Economy and Society,* 10(4), pp. 452–466.

Tucker, G. L. S. 1954. "The Origin of Ricardo's Theory of Profits." *Economica,* 21, pp. 320–33.

______ 1976. *William Huskisson Essays of Political Economy.* Canberra, Australia: Australian National University.

Wadsworth, A. P., and Mann, Julia. 1965. *The Cotton Trade and Industrial Lancashire 1600–1780.* Manchester: University of Manchester Press.

Wall, C. E. (Ed.). 1971. *Poole's Index to Periodical Literature 1802–1906.* 6 Volumes, Ann Arbor: Pierian Press.

Wallas, G. 1898. *The Life of Francis Place, 1771–1854.* London: Longmans, Green.

Ward, J. T. 1970. *Popular Movements 1830–1850.* London: MacMillan.

Wardell, M., and Turner, S. (Eds.). 1986. *Sociological Theory in Transition.* Boston: Allen and Unwin.

Weatherall, D. 1976. *David Ricardo. A Biography.* The Hague: Martinus Nijhoff.

Weber, M. 1958. *The Protestant Ethic and the Spirit of Capitalism.* New York: Charles Scribner's Sons.

Wheatley, J. 1803. *Remarks on Currency and Commerce.* London: Cadell and Davies.

_____ 1807. *An Essay on the Theory of Money and Principles of Commerce.* London: Cadell and Davies.

Whitley, R. 1984. *The Intellectual and Social Organisation of the Sciences.* Oxford: Clarendon Press.

Willer, D and J. 1971. *Systematic Empiricism: Critique of a pseudo-science.* New York: Prentice Hall.

Winch, D. 1966. *James Mill, Selected Economic Writings.* Edinburgh and London: Oliver and Boyd.

_____ 1972. *Economics as Policy: A Historical Survey.* England: Collins/Fontana.

_____ 1978. *Adam Smith's Politics: an essay in historiographic revision.* Cambridge: Cambridge University Press.

Winch, P. 1959. *The Idea of Social Science, and Its Relations to Philosophy.* London: Routledge and Kegan Paul.

Wiseman, J. 1980. "The Sociology of Knowledge as a Tool for Research into the History of Economic Thought." *American Journal of Economics and Sociology,* 39, pp. 83–94.

_____ (Ed.). 1983. *Beyond Positive Economics?* New York: St. Martin's Press.

Wolff, R. D., Roberts, B., and Gallari, A. 1982. "Marx's (Not Ricardo's) Transformation Problem: A Radical Reconceptualisation." *History of Political Economy,* 14(4), pp. 564–82.

Woolgar, S. 1981. "Interests and Explanation in the Social Study of Science." *Social Studies of Science,* 11(3), pp. 365–94.

Wynne, B. 1983. Technology as Cultural Process. Working Paper. wp–83–118. *International Institute for Applied Systems Analysis.* Laxenburg, Austria.

Yearley, S. 1982. "The Relation between Epistemological and Sociological Cognitive Interests: Some Ambiguities Underlying the use of Interest Theory in the Study of Scientific Knowledge." *Studies in the History and Philosophy of Science,* 13(4), pp. 353–88.

_____ 1985. "Vocabularies of Freedom and Resentment: A Strawsonian Perspective on the nature of Argumentation in Science and the Law." *Social Studies of Science,* 15, pp. 99–126.

Zahler, R. S., and Sussmann, H. J. 1977. "Claims and Accomplishments of Applied Catastrophe Theory." *Nature,* 296, pp. 759–63.
Zeeman, E. C. 1971. "The Geometry of Catastrophe." *Times Literary Supplement,* 10th December 1971, London: Times.

CONTEMPORARY PERIODICALS

Aris's Birmingham Gazette.
Blackwood's Edinburgh Magazine.
The Black Dwarf.
The Edinburgh Review, or Critical Journal.
The Gorgan.
The Monthly Review, or Literary Journal.
The Morning Chronicle.
The Political Register.
The Quarterly Review.
The Times.
The Westminster Review.

Index